The Wildlife of Costa Rica

# The Wildlife of Costa Rica

## A FIELD GUIDE

**FIONA A. REID**
**TWAN LEENDERS**
**JIM ZOOK**
**ROBERT DEAN**

**A Zona Tropical Publication**

FROM

COMSTOCK PUBLISHING ASSOCIATES
*a division of*
Cornell University Press
Ithaca, New York

Mammal text, arthropod text, sidebars, and introduction by Fiona Reid; reptile text and amphibian text by Twan Leenders; bird text by Jim Zook; mammal illustrations by Fiona Reid; bird illustrations by Robert Dean; reptile, amphibian, and arthropod illustrations by Twan Leenders.

The Fiona Reid illustrations on the following pages were first published in *Field Guide to the Mammals of Central America and Southeast Mexico* by Reid, Fiona (1998), and appear here by permission of Oxford University Press, Inc.: 2 (top), 4 (top), 5, 6, 8, 9, 10, 12 (top), 13, 14, 15, 16, 17, 18, 19, 20, 21, 22, 23, 24, 25, 26, 28, 29, 30, 31, 32, 33, 34, 35, 36, 38, 39, 41 (bottom), 42, 43, 44, 45.

Three Fiona Reid illustrations used in this book first appeared in a new edition of *Field Guide to the Mammals of Central America and Southeast Mexico* by Reid, Fiona (2009), and are reproduced here by permission of Oxford University Press, Inc.: 3 (top), 4 (bottom), 41 (top).

The photographers maintain copyright on their respective images:

Michael and Patricia Fogden: xi (bottom), 37, 40, 46, 57, 127, 173, 189, 202, 206, 207, 215, 224, 227; Adrian Hepworth: xii (top), xiii (top), xiv, opposite p. 1, 27, 155, 163, 195, 233; Kenji Nishida: 222, 235, 237; Roy Toft: viii, ix, xi (top), xii (bottom) xiii (bottom), 7, 11, 92, 145, 158, 198, 211, 247.

First published 2010 by Cornell University Press
First printing Cornell Paperbacks, 2010

*Library of Congress Cataloging-in-Publication Data*

The wildlife of Costa Rica : a field guide / Fiona A. Reid ... [et al.].
    p. cm.
"A Zona Tropical publication."
Includes bibliographical references and index.
ISBN 978-0-8014-4905-5 (cloth : alk. paper) -- ISBN 978-0-8014-7610-5 (pbk. : alk. paper)
1. Animals--Costa Rica--Identification. I. Reid, Fiona, 1955- II. Title.
QL228.C8W55 2010
591.97286--dc22                      2009044407

Zona Tropical ISBN: 978-0-9816028-1-3 (pbk.: alk. paper)

Printed in China
Cloth printing          10 9 8 7 6 5 4 3 2 1
Paperback printing    10 9 8 7 6 5 4 3 2 1

Book design: Zona Creativa S.A.
Designer: Gabriela Wattson

# Contents

# List of Natural History Vignettes

In addition to the species descriptions in the pages that follow, this book includes a series of natural history vignettes—each consisting of a photograph and a text nugget—that highlight aspects of the animals of Costa Rica, their fascinating habits and complex relationships to the environment. These invite you to further explore the beautiful and wonderfully complex world of Costa Rican wildlife.

# Acknowledgments

**Fiona A. Reid** would like to thank the following people for help with aspects of the mammal text: Judith Eger, Mark Engstrom, Kris Helgen, Burton Lim, Sandra Peters, Bernal Rodríguez, Jan Schipper, and Rob Voss. She received much help with the arthropod text from Chris Darling, Paul Hanson, and Kenji Nishida. All written material was edited by John McCuen, whose efforts were indispensable in unifying the work of several different authors. Finally, Fiona would like to thank her husband, Mark Engstrom, and children, Holly and Ian Engstrom, for their support.

**Twan Leenders** would like to thank Tim Paine, Tobias Eisenberg, the folks at Rara Avis and the Rainmaker Conservation Project, as well as all the staff and students of the Forman School Rainforest Project for their support during many years of field work. The encouragement (and eye for detail) provided by John McCuen and Marc Roegiers from Zona Tropical were instrumental in the production of this book. Most important, he would like to thank his wife, Casey, and kids, Madeleine and Jason, for giving him the opportunity to work on this project.

**Jim Zook** would like to thank parents Bill and Barbara Zook for cultivating his interest in nature and birds from an early age. He also thanks his wife, Berny, and son, Anthony, for their support and encouragement during this project. Finally, he notes his deep appreciation for Gretchen Daily and Paul Ehrlich, who kept him in the field, and close to the action, these many years.

**Robert Dean** wishes to thank the following people for having provided assistance in seeing birds he'd not seen before: Eduardo Amengual, Marino Chacón, Judy Davis, Kathy Erb, David Fisher, Michael and Patricia Fogden, Aisling French, Tony Godfrey, Larry Landstrom, Jamie Midgley, Pete Morris, Patrick O'Donnell, Alison and Michael Olivieri, Steve Pryor, Candy and Rich Stewart, Alex Villegas, Mark Wainwright, Roberto Wesson, and Jim Zook. He also thanks Julio Sánchez and the Natural History Department at the Museo Nacional de Costa Rica for having graciously provided access to the bird collection. And lastly, he expresses his gratitude to Richard Garrigues for the fine work he contributed to *The Birds of Costa Rica: A Field Guide*.

# Introduction

Costa Rica, a peaceful nation with many and diverse animal species, is one of the best places in the world for wildlife watching and nature study. It has an excellent system of national parks and reserves, a wide choice of eco-lodges, and a cadre of professionally trained tourist guides. In a country no bigger than West Virginia, it is possible to leave the capital city of San José and find oneself, just a few hours later, in a high-elevation cloud forest, dense rainforest, savanna-like plain, or coastal habitat, each with a unique collection of animal species.

There is an ever-growing list of publications about the natural history of Costa Rica, ranging from scientific tomes such as Savage's *Reptiles and Amphibians of Costa Rica* to laminated foldouts. In this book, we hope to provide a general presentation of the wildlife of Costa Rica that is at once lightweight and affordable, that provides details for identifying animals along with interesting facts about their natural history, and that also offers some tips for seeing them. This is not a guide to all the wildlife in Costa Rica—indeed, such a guide would be encyclopedic in length and require an infinity of time to prepare—but it will serve as a useful, informative introduction to the animals seen on a walk through the forest.

Selecting which species to include in the book—of all the mammals, birds, reptiles, amphibians, and arthropods in Costa Rica—represented an interesting challenge. The majority of the species were selected because they are animals that are relatively easy to see and either beautiful or interesting in some other way (or both). A small group of charismatic or distinctive animals—the Jaguar and the Bushmaster, for example—were included even though you are unlikely to see them. It was thought that presenting some of these rare or secretive species would give a fuller picture of the amazing diversity of wildlife in the country. Plus, people want to know about these animals, even if the chances for seeing them are slim.

The percentage of species covered varies from group to group. The bird chapter, for example, includes accounts for a significant percentage of the total species in the country. For arthropods, on the other hand, we include accounts for only a small percentage of the many, many thousands of species. In deciding how many species to include for each group, we took into account two principal factors. First, was an assessment of the level of interest on the part of most readers. Birds and mammals are high on the list of many nature enthusiasts—or so we gauged it—and they are, accordingly, given relatively thorough coverage. The second factor relates to the ease—or difficulty—of identifying an animal (or group of animals) in the field. Bats, many species of frog, colubrid snakes, and many arthropods are

Brown-throated Three-toed Sloth (*Bradypus variegatus*).

Scarlet Macaws squabble over food.

often well-nigh impossible to distinguish one from the other in anything other than laboratory conditions. In general, then, we tended to avoid including species that are difficult to identify; when they are included, however, they serve to represent a collection of animals (a genus or a family) rather than a specific species.

   *The Wildlife of Costa Rica* is the first ever such guide to combine color illustrations, a wealth of natural history information, identification traits, and treatment of all the major phyla in the country. We think you will find it an indispensable companion on your next walk through the rainforest and hope you will enjoy it. The wildlife of Costa Rica awaits you!

## Geography and Habitats

Some two to seven million years ago, a section of land encompassing what today are Costa Rica and Panama gradually rose from the ocean to connect North and South America. The two continents had existed apart for over 60 million years, during which time a very distinct fauna and flora had evolved on each. With the formation of what is known today as the Panamanian land bridge, animals began to move from north to south and from south to north, and this resulted in an increasingly diverse composition of species in Central America. Costa Rica is consequently home to species that evolved in North America, those that evolved in South America, and, in a third category, endemic species.

   Bounded by the Caribbean Sea and the Pacific Ocean, Costa Rica is a small country oriented along a northwest to southeast axis. Four mountain chains run down the center of the country—the Guanacaste, Tilarán, Central, and Talamanca ranges—among them the highest peak in Central America, Mount Chirripó, and several active volcanoes. The backbone

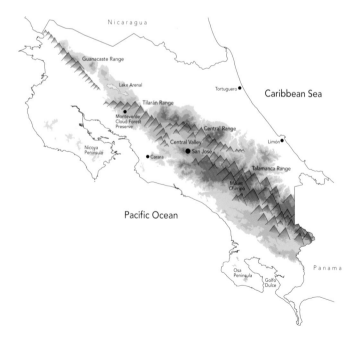

of these mountains is known as the Continental Divide. The entire region to the west of this divide is referred to as the Pacific slope, not to be confused with the term *Pacific coast*, which designates coastal areas only; similarly, to the east of the divide lies the Caribbean slope. The Central Valley, where the majority of Costa Ricans live, is a large, high-elevation valley enclosed by the Central Mountains and the northern part of the Talamanca Mountains.

The Caribbean side of the country consists of a wet coastal plain that is widest toward the Nicaraguan border and narrowest toward the southern border with Panama. Simplifying somewhat, the Pacific side of the country divides into the dry northwest region, which encompasses the lowland plains of Guanacaste and the Nicoya Peninsula, and the rainy Pacific lowlands that extend south from Carara to the Osa Peninsula.

Relative to its size, Costa Rica has more biodiversity than either of its neighbors, a fact explained in part by the large number of distinct habitats that occur within its borders. Biologists describe an astounding twelve life zones for the country, but these can be simplified somewhat and divided into the categories that follow.

RAINFOREST. Alternative labels include lowland rainforest, tropical moist forest, and wet or evergreen forest. To those from North America or Europe, this is the life-zone type that most closely conforms to that somewhat cinematic concept— The Jungle. In point of fact, several types of forest are subsumed under this category, although they all share a number of characteristics. The trees are tall, and the very tallest—the canopy trees—are green throughout the year. Rainforests are composed of distinct layers: canopy and emergent trees form the upper layer;

Rainforest on the Osa Peninsula.

subcanopy trees are somewhat smaller; then come understory trees, shrubs, and, lastly, undergrowth. Palms are common in both understory and shrub layers. Lianas and huge tree buttresses are nearly emblematic of this kind of forest. In Costa Rica, rainforest occurs in much of the Caribbean lowlands, an area with no pronounced dry season, and, on the Pacific slope, in much of the area south of Dominical, especially the Osa Peninsula.

HIGHLAND FOREST. This category includes cloud forests, montane oak forests, and elfin forests, all of which exist in Costa Rica. The term *cloud forest* refers to cool forests between 3,300 and 6,550 ft (1,000 and 2,000 m) that are perpetually enshrouded in clouds. Cloud forest trees tend to have broad leaves, may reach a height of 100 ft (30 m), and are often festooned with epiphytes. The Monteverde region is home to Costa Rica's most visited cloud forest. Montane oak forest, found at high elevations on the Central and Talamanca mountains, contains huge oaks (*Quercus* spp.) of up to 125 ft (40 m). The oaks, laden with red bromeliads, mosses, and other epiphytes, tower above an understory of bamboo, ferns, and

Monteverde cloud forest.

prehistoric-looking tree ferns. Mountain tops and ridges, including some peaks that are not at extreme elevations, may be covered by elfin forest, which consists of stunted, gnarled trees, usually only 15 ft (5 m) high; numerous mosses and ferns; and a dense layer of shrubs. Above the tree line, on the tallest of the Talamanca Mountains, is yet another habitat type, páramo, which is made up of cold, open zones with grasses, dwarf bamboo, cushion plants, and shrubs.

TROPICAL DRY FOREST. Also known as lowland tropical deciduous forest. Occurs in regions with a long, pronounced dry season, during which most trees lose their leaves

Tropical dry forest.

(some trees may flower at this time). In these forests the trees are not very tall and do not form a closed canopy. Characteristic tree species include the calabash, pochote, and emblematic Guanacaste tree, a tall, spreading giant. Woody vines are common, but other vines and epiphytes are sparse. In Costa Rica, tropical dry forest occurs in the northwest region of the country, where it is punctuated with patches of evergreen forest along rivers (gallery forest) and in bottomlands. Much of the original dry forest has been converted to pasture for cattle ranching.

Mangrove on Pacific coast.

MANGROVES. In Costa Rica, mangroves are found mainly on the Pacific coast and on the northern Caribbean coast. Mangroves consist of a few tree species (*Rhizophora* sp. and others) with aerial roots that allow the trees to survive in soil inundated by salt water.

SWAMPS and MARSHES. Swamps are flooded areas in forests; marshes are wet areas in open habitats. Large areas of swamp forest—often dominated by *Raphia* palms—occur in the northern Caribbean lowlands of Costa Rica. These areas have black waters resulting from slowly decaying leaves. There are important marshland habitats at Palo Verde, in the northwest Pacific, and at Caño Negro, near Arenal Volcano.

Black-bellied Whistling-Ducks (*Dendrocygna autumnalis*) taking flight at Palo Verde.

DISTURBED HABITATS. This is an umbrella term used to describe any habitat that—whether through human agency or natural forces—has been altered from its original state. This includes disturbed forest, open woodland, savanna and grassland, agricultural areas, and gardens and urban parks. Ecological concerns aside, disturbed habitats often afford opportunities for bird and animal watching. Note that the term *disturbed forest* is defined in relation to the concept of mature (or primary forest), which is any tall, pristine forest with an unbroken canopy. Secondary forest is forest that was once logged or depleted through natural causes, but has begun to regrow. Even in mature forest, however, the habitat along trails often consists of disturbed vegetation. Depending on the age of a secondary forest, the trees may actually be quite tall; given enough time, secondary forest will eventually become a mature forest.

Deforested terrain in southern Costa Rica.

## Weather

Costa Rica is located within the tropics and experiences just two seasons—a dry season and a wet season. The dry season usually begins toward the end of November and lasts until April or May, when the rainy season begins; October is the rainiest month of the year. However, the weather does vary significantly from region to region; during the dry season, it often continues to rain on the Caribbean slope. And in September and October, usually the two rainiest months of the year in most of the country, the Caribbean coast—especially the southern Caribbean coast—experiences clear skies, and the Caribbean Sea transforms itself into a large, placid lake—at least in the vicinity of Costa Rican shores.

Yellow cortez (*Tabebuia ochracea*) in full bloom in the dry season, Guanacaste National Park.

The rainiest low-elevation regions— the ones tourists are likely to visit, anyway—are the northern Caribbean and the southern Pacific, in the area of Golfo Dulce and the Osa Peninsula. On the other end of the spectrum is the hot northwest Pacific (including the Nicoya Peninsula), where the dry season is harsh and long—it sometimes lasts up to eight months. At elevations above 4,900 ft (1,500 m), the temperature becomes cooler and, higher up, rain gives way to clouds and mist. Near the top of some of the highest peaks, there is occasional frost.

## A Note on Taxonomy

Intended as it is for the general reader, this book focuses on the species level and concerns itself with detailing family or order characteristics only when such information might explain how diverse a given family is, provide important context for the species descriptions, give some interesting natural-history tidbit, or in some other way enlighten the reader. In many cases, information about the family is included within the order description, as in this particularly concise example:

### Order Lagomorpha (Rabbits)

Lagomorphs are found worldwide. There are three species in Costa Rica, all in the family Leporidae.

At the start of most chapters mention is made of what taxonomic convention the author has adhered to in presenting the animals in that chapter. By way of example, in the bird chapter, the names and taxonomy follow the Checklist of North American Birds (8th supplement to the 7th edition), published by the American Ornithologist's Union.

For those interested, A Note on Amphibian Taxonomy, p. 252, indicates some of the maddeningly complex taxonomic issues that confront biologists who study Costa Rican frogs and toads.

## About This Book

The five chapters of this book cover mammals, birds, reptiles, amphibians, and arthropods. Each starts off with a brief introduction that cites the number of species that occur in Costa Rica, mentions some of the group's more interesting representatives, and gives some tips for seeing them. For the most part, species accounts all follow the same format. A physical

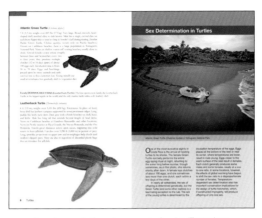

Sample spread of species accounts showing illustrations
and natural history vignette.

description of the animal, including body measurements and, sometimes, a description of similar-looking species, is followed by information about range—usually indicating both elevation and habitat preferences—and, finally, details about the natural history of the species. It is accompanied by a color illustration of the animal and sometimes an additional illustration or two to indicate variation with age and sex. Within each two-page spread, illustrations are to scale except when indicated by the notation *not to scale*, in which case you can refer to the written measurements to determine relative size.

Note that citations for body measurements were originally stated in metric then translated into inches and feet, and rounded off with increasing precision for smaller figures. When discussing elevation ranges for animals, some accounts cite feet/meters figures whereas other accounts rely simply on a text description—e.g. "found at low and middle elevations." The general rule of thumb was to use numbers where hard data were available, but go with general statements of elevation range when hard data were not at hand or, in the case of some species, when animals range so freely from elevation to elevation that numbers would suggest a precision that doesn't exist in reality.

Interspersed throughout the text is a series of natural history vignettes, each consisting of a photograph and text that together portray some general point about ecology or natural history or that describe in greater detail a given species. Several species described in these vignettes do not appear within the species accounts—they are extras—but you will find them included in the index of species names. At the back of the book there is a glossary and a brief bibliography with suggestions for readers who want to dig deeper in specific areas. At the very end of the book, you will find the index to scientific and common names to help you locate the species account for a specific animal.

mammals

**Costa Rica's** postage-stamp terrain is home to an astonishing number and diversity of mammals. Indeed, its approximately 110 species of bats is about two and a half times the number known from the United States and Canada combined. And as advanced DNA sequencing techniques are applied to the study of bats, shrews, rodents, and small opossums, biologists continue to identify new species. The current mammal count for Costa Rica stands somewhere between 240 and 250 species, with a more exact tally expected as taxonomic controversies are resolved.

The isthmus of Central America is quite literally a land bridge that rose from the oceans several million years ago to connect the northern and southern continents. The movement of North American mammals southward—and of South American mammals in the opposite direction—along with the variety of habitats and climate zones found in Costa Rica, all led to the diversity of wildlife that occurs in the country today. Some species—the White-tailed Deer, for example—will be familiar to many visitors to the country. A few groups of mammals, including some of the arboreal rodents, evolved in Central America and have no close affiliation with species in either North America or South America. But visitors to the country, by and large, want to see the charismatic mammals that have come to symbolize the tropical world, among them the monkeys, cats, kinkajou, tapir, sloths, anteaters, and manatee.

This chapter includes a mix of mammals that you are likely to see and mammals that you might hope to see (but might not have the good fortune to see). Thus, only a small sample of bat species are included since the majority are seldom seen and are extremely difficult to identify without capture. All six of the cat species that occur in Costa Rica are described, although they are almost never seen. Mammals are often difficult to observe in the wild; many are nocturnal and larger mammals tend to occur in smaller numbers, even in pristine habitats.

In Costa Rica, however, it is possible to see a variety of mammals by following a few suggestions. If you spend time in some of the large, more remote national parks (Tortuguero and Corcovado, for example), you are virtually guaranteed to see monkeys and other mammal species. Manuel Antonio National Park, though rather small, affords ample opportunity to see sloths, monkeys, and opossums. During the dry season, it's worth visiting Palo Verde and Santa Rosa national parks. At waterholes there, patient observers can see the mammals that come to drink in the afternoon—and at dawn and dusk. At night, use a flashlight and binoculars to look for eyeshine from nocturnal creatures. Thus equipped, you can often see mammals just a short distance from your lodging.

Baird's Tapir (*Tapirus bairdii*). This tapir, the largest land mammal in Costa Rica, can move through dense forest with surprising ease and silence. It uses its excellent sense of smell to avoid detection and to find food.

Mammal taxonomy and order of the species accounts conform to Wilson and Reeder (2005), with a few exceptions in consideration of page design. Unless otherwise indicated, measurements are for average total length of an adult, from tip of nose to tip of tail. This does not include long hair projecting beyond the tail bone.

---

**Order Didelphimorphia (American Opossums)**
All marsupials were formerly classified in the order Marsupialia, but this large group was split recently into seven orders. The Didelphimorphia contains a single family (Didelphidae), which includes 8 species (possibly 9) in Costa Rica (91 species throughout the New World). The tail is usually long, naked, and prehensile. A well-developed pouch is present in the larger species; the pouch is absent in mouse opossums. Costa Rican opossums are nocturnal. All species are solitary outside the breeding season. Leaves and grasses are used to build nests; these materials are carried in coils of the tail. Gestation is short; after birth the tiny young crawl unassisted to the mother's nipples, where they attach themselves for several weeks or months.

---

## Mexican Mouse Opossum (*Marmosa mexicana*)

10 in (27 cm). About the size of a large mouse, but easily distinguished by its large eyes and black eye rings. Note also the opposable thumbs on hind feet. Fur bright orange or pale orange above, cream below. Secretive and rarely seen, though it is quite common throughout CR. Found at all levels of the forest, from understory to canopy, but seldom comes to the ground. Clings with its long naked tail as it moves among vines and branches. Eats insects and fruit. Sometimes constructs its ball-shaped nest inside boxes erected for bird nests.

## Alston's Mouse Opossum (*Micoureus alstoni*)

17 in (43 cm). Fur woolly and grayish above, pale yellow or cream below. Prominent black eye ring. Tail is furred for about 2 in (5 cm) at its base. This is the only small opossum with a white-tipped tail. It is slightly larger than the Mexican Mouse Opossum. Relatively uncommon; occurs on the Caribbean slope, at low and middle elevations, and in the Central Valley. Although most opossums are solitary, this species sometimes lives in groups. It may invade houses near forested areas.

## Central American Woolly Opossum (*Caluromys derbianus*)

25 in (64 cm). Fur gray with variable amounts of bright orange on shoulders and rump. Underparts pure white. Ears large, pale pink. Tail thickly furred for half its length, with a long, bare, white tip. Common; occurs in forested areas throughout the country. Strictly arboreal and nocturnal; it may be encountered at night feeding in balsa and cecropia trees, its large eyes reflecting a bright orange-red color. Sometimes seen crossing roads on power lines or phone lines, but almost never descends to ground. Very fast and agile, it uses its long tail for balance when climbing. Feeds on fruit, nectar, and insects. Makes leaf-nest in vine tangles and in tree holes far above ground.

## Gray Four-eyed Opossum (*Philander opossum*)

23 in (58 cm). A pair of white spots above the eyes gives the impression of four eyes, hence the common name. A handsome dark-gray opossum. Tail furred at base; midsection naked and black; long white tip. Common; occurs throughout CR at low and middle elevations. Nocturnal and semi-arboreal. Most often found near small streams and waterways in forested areas. Omnivorous; eats small vertebrates, invertebrates (including crabs, shrimps, and insects), and fruit. Constructs leaf-nest within vines or in hollow trees. Females have litters of 2 to 7 young and may breed two times per year.

## Common Opossum (*Didelphis marsupialis*)

27 in (70 cm). Scruffy fur and naked tail convey the impression of an oversized house rat. Body grizzled, yellow gray; cheeks yellowish. Tail naked, black at base; long white tip. (Virginia Opossum, in N.W. Pacific only, has white cheeks and a shorter tail.) Common, occurs throughout CR in forests and at forest edges. At night it may be seen on the ground or on a tree branch; freezes if illuminated, staring at observer. If threatened it rocks from side to side then hisses and twirls around, spraying urine or feces. This species does not "play dead." Eats almost anything, including carrion, and will raid hen houses. Dens in hollow trees or vine tangles. Litters can be as large as 20 young, but only about 6 survive in pouch.

## Water Opossum (*Chironectes minimus*)

26 in (66 cm). Also known as Yapok. Pale gray above; four broad blackish bands across back; belly a uniform white. Hind feet webbed. Tail broad, mostly naked and black, with a short white tip. Seldom seen; occurs on Caribbean and Pacific slopes at low and middle elevations. Semi-aquatic and always found close to water. Uses large hands to grab fish and invertebrates in small rivers and streams. Constructs den in banks. Litters of 2 to 5 young remain in the pouch when the mother hunts. The female's pouch is highly muscular and forms a watertight compartment for the young when closed.

### Order Pilosa (Sloths and Anteaters)

Sloths and anteaters, though externally dissimilar, share certain skeletal characteristics. (Both were formerly included with armadillos in the order Xenarthra.) These mammals diverged early from the main placental mammal lineage, when South America was an island continent.

**Family MEGALONICHIDAE (Two-toed Sloths).** A single species from this family occurs in Costa Rica; another species is restricted to South America. These sloths are allied with an extinct group of giant ground sloths.

## Hoffmann's Two-toed Sloth (*Choloepus hoffmanni*)

20 in (50 cm). Fur long, pale brown; darker on limbs, whitish on forehead. Broad, piglike snout. Two claws on front foot, three claws on hind foot. No tail. Young is dark brown. All four limbs are about the same length. Fairly common; occurs in wet forests on Caribbean and Pacific slopes (absent in dry N.W. Pacific). Found at low and middle elevations (Arenal, Monteverde, and Tortuguero, for example), but also occurs in highlands. Suns itself on an exposed branch in the morning; sometimes active by day but usually feeds at night. Faster moving than Three-toed Sloth; it can support its weight and crawl on the ground. Both sloths give high-pitched cries or whistles, presumably in territorial disputes. *Illustration not to scale.*

**Family BRADYPODIDAE (Three-toed Sloths).** There is one species of three-toed sloth in Costa Rica. A new species from western Panama, the Pygmy Three-toed Sloth, was recently described, and two other species occur in South America. Interestingly, the four species in this family are not closely related to the two-toed sloths.

### Brown-throated Three-toed Sloth (*Bradypus variegatus*)

24 in (60 cm). Fur coarse, grizzled, gray. Small dark snout; black mask through eyes; seems to wear a perpetual smile. Limbs each bear three long claws. Short tail. Male has a patch of orange and black fur on back. Arms longer than legs. Common to abundant in Manuel Antonio, Corcovado, Tortuguero, and Cahuita national parks. Often seen sunning itself in cecropia trees. Eats leaves of this tree and various other tree species. Mainly active at night, but sometimes feeds during the day. Cannot support its weight on the ground, but swims proficiently. Descends to the ground once a week to defecate in a small hole it digs with tail. Local common name of both Costa Rican sloths is *perezoso* (the lazy one).

# Brown-throated Three-toed Sloths

Male Brown-throated Three-toed Sloth (*Bradypus variegatus*). The three-toed sloths are thought to make up the largest mammal biomass in Costa Rica, humans aside.

Three-toeds' green-tinged fur makes them very difficult to spot when they are curled up asleep in the crotch of a tree. The green color derives from algae that live inside the fur; each hair is a hollow tube slit lengthwise, a structure that allows the algae access to the interior of the tube. Sloth moths that inhabit the fur keep algal growth in check by eating the algae.

Sloths eat leaves, an extremely low-energy diet. By necessity, then, they move slowly to conserve limited reserves of energy. Males advertise their presence to other males by displaying an orange and black patch on the midback and by whistling, two methods of establishing territorial rights that are much more energy efficient than active patrolling.

Why does this calorie-starved creature make weekly visits to the forest floor to defecate when it could do the same from its tree perch—without expending energy? There are two popular theories, both plausible. One line of thought is that sloths provide fertilizer for the host tree when they deposit their droppings in a small hole at its base. Another explanation involves sloth moths; they fly off the sloth to lay their eggs in sloth feces and then return to the sloth before it makes its way back up the tree. Thus the moths have easy access to a good site for larval development without having to abandon their host.

**Family MYRMECOPHAGIDAE (Anteaters).** This family is represented by a single species in Costa Rica (a closely related species is found in South America). There is some evidence, however, that the rare Giant Anteater also occurs in the country. Anteaters were formerly included in the same order as armadillos, but are now grouped only with sloths. Anteaters have long noses, a wormlike tongue, and no teeth.

## Northern Tamandua (*Tamandua mexicana*)

42 in (110 cm). Medium size. Mostly golden, with a contrasting black vest (some individuals entirely golden). Prehensile tail. Relatively common; occurs throughout CR in a variety of habitats, but is most easily seen in dry forests of the N.W. Pacific. Diurnal and nocturnal. Semi-arboreal; travels on the ground and through trees in search of termite and ant mounds. Sometimes located by the tearing sounds it makes as it rips into rotten wood. When threatened, it sits up, using its tail as a brace, and strikes with massive foreclaws. It is persecuted by local farmers, perhaps because it can easily kill an inquisitive dog. Local name is *oso hormiguero*.

**Family CYCLOPEDIDAE (Silky Anteater).** The single member of this family is widespread in Central America and South America, though it is seldom seen and poorly known.

## Silky Anteater (*Cyclopes didactylus*)

15 in (38 cm). Tiny. Woolly, silvery gold fur; prehensile tail; two large, sharp claws on forefoot. Occurs at low and middle elevations; though generally rare (and seldom seen), it is common in several mangrove swamps, including the one that borders Damas Island, just off the coast of Quepos. Arboreal and nocturnal. During the day, curls into a furry golden ball and rests in vine tangles or on narrow branches. At night, travels along thin vines and branches in search of ants and termites. Rips opens hollow branches with its foreclaws and extracts ants with a sticky tongue. Local Spanish name *tapacara* (covered face) derives from defensive strategy of shielding its face with powerful foreclaws.

**Order Cingulata (Armadillos)**
There are two species of armadillo in Costa Rica, both in the family Dasypodidae. The Northern Naked-tailed Armadillo is rare and poorly known; the Nine-banded Armadillo, also known as the Texas Armadillo, has a range that extends from the United States to southern South America.

## Nine-banded Armadillo *(Dasypus novemcinctus)*

33 in (83 cm). Armor encases both domed shell and long tail; 8 to 9 bands on the back; ears, long and narrow, are placed close together. (The Northern Naked-tailed Armadillo has a flatter shell, no armor on the tale, and broad, widely spaced ears.) Common and widespread; occurs throughout CR in a variety of habitats, and may be active by day or night. Eats invertebrates, fruit, and carrion; sometimes located by the rustling and snuffling sounds it makes as it rummages through the leaf litter. An agile swimmer; crosses streams either by walking along the bottom or by gulping air and then floating across. These armadillos mate face-to-face. Females give birth to quadruplets; because they are genetically identical (and susceptible to human diseases), the quadruplets have been used to research leprosy and other maladies. *Illustration not to scale.*

### Order Primates (Monkeys)

This order includes the lemurs, tarsiers, and bushbabies, as well as the more familiar monkeys and apes. The majority of monkey species occur in tropical regions. There are 4 primates in Costa Rica, all arboreal and all threatened by the effects of deforestation.

**Family CEBIDAE (Squirrel and Capuchin Monkeys).** This family also includes the marmosets and tamarins, found in Panama and South America. The Red-backed Squirrel Monkey, one of two cebids in Costa Rica, is the smallest primate in the country.

## Red-backed Squirrel Monkey (*Saimiri oerstedii*)

Osa race

27 in (68 cm). Small, colorful, and lively. Cap, muzzle, and tip of tail either black or dark gray; back and lower legs orange brown; shoulders, thighs, and tail olive brown. Northern race (north of Térraba River) is paler and grayer than southern race. Endangered; occurs only in Manuel Antonio area (northern race) and on the Osa Peninsula (southern race). Prefers tall secondary forest. Groups of 10 to 40 travel rapidly and noisily through the forest, investigating every nook and cranny in search of insects and small vertebrates. Also eats fruit and nectar in the dry season. The *mono tití*, as it is called in CR, suffers from habitat loss, illegal captures for the pet trade, and heavy pesticide use.

## White-faced Capuchin (*Cebus capucinus*)

36 in (90 cm). Medium size. Mostly black, with cream on head, chest, and shoulders; face pink. Thick, prehensile tail is often coiled. Found throughout the country in all forest types, including mangrove and highland forest. Fairly common in protected forests, but may be hunted in unprotected areas. Easily seen in Palo Verde and Santa Rosa national parks and in the Tortuguero area. This is the only primate in CR that sometimes travels and forages on the ground. Eats fruit, buds, flowers, insects, small vertebrates, and bird eggs. Lives in groups of 5 to 30 individuals; larger males defend group from predators and intruders.

# The Crafty Capuchin

White-faced Capuchin (*Cebus capucinus*).

Most primates are highly social animals, and the White-faced Capuchin is no exception. Males work in concert to defend their troop against intruders. When hunting larger prey, several troop members will band together to capture squirrels or to raid the nests of birds and coatis.

White-faced Capuchins are also intelligent. Like the great apes, they are tool users, picking up sticks to hurl at snakes or human intruders. In one instance, a capuchin used a heavy stick to beat a snake to death. These sometimes aggressive primates also collect medicinal plants—including citrus fruits, *Piper*, *Clematis*, and *Dieffenbachia*—to rub into their fur. They are known to contain chemicals that function as insect repellents, antiseptics, or fungicides. The monkeys appear to know the appropriate time to use the plants, rubbing themselves most often in the rainy season, when they suffer more insect bites and skin diseases.

**Family ATELIDAE (Howler and Spider Monkeys).** Two members of this family occur in Costa Rica. Both species have prehensile tails and an arboreal lifestyle. Howler monkeys are slow-moving folivores, while spider monkeys are active fruit eaters.

## Mantled Howler Monkey (*Alouatta palliata*)

42 in (1.1 m). Large and stocky. Mainly black, with a yellow-brown mantle of long fur on the sides of the body. Prehensile tail. Male is larger than female and has a longer beard and a prominent, white scrotum. Common and widespread; occurs throughout CR in most types of forest and in forest remnants. Groups of 10 to 20 spend much of the day resting on large tree branches. When active, travels slowly, moving on all fours with the tail coiled and held low. Feeds mainly on leaves but also eats fruit, flowers, and buds. Males produce loud, deep roars, mainly at dawn and dusk. They also roar in response to thunder and other loud sounds. These sounds can travel several miles and help groups maintain proper distance from each other. Roar of female and young is softer and higher pitched than that of male.

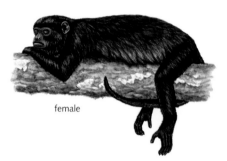

female

male

## Central American Spider Monkey (*Ateles geoffroyi*)

42 in (1.1 m). Large and long limbed. Back is usually reddish brown, contrasting with darker head, tail, and limbs; color varies from one race to another. Male and female are similar, though female has a penis-like clitoris (male genitals are usually hidden from view). Uncommon to rare; some races are endangered. Occurs throughout the country, mostly in large tracts of evergreen forest but also in dry forest, near rivers. Ten to 40 individuals sleep together in large trees; during the day, the group splits into smaller bands that forage for fruit. This monkey moves with great speed and agility, either swinging arm over arm or moving on all fours with tail held high. When alarmed, lone animals sometimes sit down and yap, the call somewhat similar to that of a Keel-billed Toucan. This species is absent from many forests due to overhunting.

**Family SCIURIDAE (Squirrels).** Worldwide there are about 280 species of squirrel. Five species occur in Costa Rica; the following accounts describe all but Deppe's Squirrel (*Sciurus deppei*), which occurs only in the Guanacaste Mountains. In addition to the familiar tree squirrels, other members of this family are the ground squirrels, chipmunks, prairie dogs, marmots, and flying squirrels. Tree squirrels are among Costa Rica's most easily seen mammals; although these diurnal creatures spend most of their time in trees, they sometimes descend to the ground to feed, and several species frequent gardens and roadsides.

## Variegated Squirrel (*Sciurus variegatoides*)

24 in (60 cm). This is the largest and most colorful tree squirrel in CR. Color is highly variable, but in almost all races the tail is frosted with white. Upperparts are commonly a grizzled gray, sometimes with a dark line on the middle of the back and a whitish line on sides; belly orange or white. Race in Manuel Antonio area is dark brown to black. Another race is entirely white with a black line along spine. Common and widespread throughout CR. Found in habitats with some tree cover; rare in mature rainforest. Sometimes seen sunning itself on branches and fencerows or foraging for fruit on the ground. Builds a large leaf-nest. *Illustration not to scale.*

### Alfaro's Pygmy Squirrel (*Microsciurus alfari*)

9 in (23 cm). Very small. Upperparts dark brown, belly white or gray. Has a short, scraggy tail. Ears very short, almost entirely concealed by fur. Uncommon; occurs throughout most of country but absent in N.W. Pacific. Found in mature evergreen forests. Travels alone or in pairs. This agile squirrel races through trees and occasionally dashes across the ground. It may be seen chewing on the live bark of *Inga* trees, from which it obtains sap; also eats fruit and insects. Usually silent, though it does make high-pitched, birdlike trills and squeaky chatters.

### Montane Squirrel (*Syntheosciurus brochus*)

12 in (31 cm). Fairly small and stocky. Tail short and bushy; ears very short, almost hidden in thick fur; belly orange. In CR, known only from Poás Volcano and Tapantí National Park, from about 6,560 to 8,200 ft (2,000 to 2,500 m). Endemic to Costa Rica and Panama. Favors wet forests and second growth. Occurs at all levels of forest and may travel on the ground; usually feeds in trees. Active during the warmest time of the day, and, in good weather, at dawn and dusk. Unlike most squirrels, it appears to form stable pair bonds and lives in family groups, although individuals may disperse to forage.

### Red-tailed Squirrel (*Sciurus granatensis*)

17 in (42 cm). Fairly large. Upperparts dark brown; belly pale to bright orange; long tail shows orange frosting. Ears project above crown of head. Common; occurs throughout CR except in the N.W. Pacific, where it is replaced by Deppe's Squirrel, a smaller species with a grayish belly. Found in evergreen and semi-deciduous forests and in second growth. Most active in the morning. Mostly arboreal but also descends to ground to forage. Can sometimes be located by rasping sounds it makes as it chews palm nuts and other hard seeds. Sleeps in hollow trees and vine tangles. Usually solitary, though large numbers of males pursue females in heat; these chases—accompanied by grunts, chucks, and squeals—can last for several hours.

**Family HETEROMYIDAE (Pocket Mice).** This family of about 60 species includes kangaroo rats and pocket mice. The taxonomy of tropical forest species is under revision. At least 4 species occur in Costa Rica, all quite similar in appearance. These rodents, and their close relatives the pocket gophers, have external, fur-lined cheek pouches.

## Salvin's Spiny Pocket Mouse (*Liomys salvini*)

12 in (31 cm). Back dark gray or gray brown; rump has fine spiny hairs; underparts white or cream. Tail long, bicolor, with a small tuft at tip. Common in N.W. Pacific (including Nicoya Peninsula); also common in Central Valley to 4,920 ft (1,500 m). Found in deciduous forest and dry, brushy second growth. With a hopping gait, searches for seeds that it collects in its cheek pouches; these pouches can be turned inside out to empty and clean. Solitary and nocturnal. During the day, occupies deep burrows that descend vertically from the forest floor.

**Family CRICETIDAE (Rats and Mice).** This family is a large, heterogeneous assemblage, with about 680 species in 130 genera. Whether to place the smaller rodents in the family Cricetidae or Muridae is an ongoing discussion among biologists; both names have been applied to Costa Rican rats and mice. Seldom seen and difficult to identify, many of the species that occur in the country have been omitted from the following pages.

## Vesper Rat (*Nyctomys sumichrasti*)

9.5 in (24 cm). Small, stocky. Short nose; large eyes; dark mask. Upperparts orange; belly white; tail long and hairy, with a pronounced tuft at tip. Uncommon through most of its range though common at some locations; occurs on Caribbean and Pacific slopes to 5,910 ft (1,800 m). This colorful rodent is strictly arboreal and very rarely captured in traps set on the ground. Feeds on fruit and insects; sometimes enters dwellings in forests, ascending to rafters in search of insects. Uses shredded bark to construct platform nests in holes in trees. Female usually gives birth to twins that are well developed at parturition. Adult emits twittering, birdlike calls during the breeding season.

## Watson's Climbing Rat (*Tylomys watsoni*)

20 in (51 cm). Large. Long, naked tail has white tip. Upperparts gray or gray brown; belly white or cream. Common; occurs throughout CR except in the dry N.W. Pacific. Endemic to Costa Rica and Panama. Occupies evergreen and semi-deciduous forests and dense second growth. Often dens in caves and rock crevices, but also takes up residence in cabins and other buildings surrounded by forest. Emerges at night to feed on fruit and vegetation, but also scavenges soap and human foods (it is especially fond of chocolate). Travels equally adeptly on the ground and on vines and branches.

**Hispid Cotton Rat** (*Sigmodon hirsutus*)

10 in (26 cm). Moderately large and stocky. Volelike. Long, grizzled gray-brown fur. Tail shorter than head and body; snout short; ears broad and rounded. Common to abundant on Pacific slope; uncommon in Caribbean lowlands. Found in grassland, brush, and swamps. Active by day, this is the rat most likely seen dashing across roads or moving through runways in tall grass. Eats green vegetation and fungi; leaves clippings of grass in active runways; sometimes a pest of rice, sugar cane, and other crops. Makes spherical nests in tall grass or under logs. Female produces litters of up to 14 young. (Formerly *Sigmodon hispidus*.)

**Alston's Singing Mouse** (*Scotinomys teguina*)

4.5 in (11 cm). Tiny. Tail about three-fourths the length of head and body. Upperparts chocolate brown; belly pale orange; feet and tail dark. Common to abundant; occurs at middle and high elevations, from 2,950 to 9,510 ft (900 to 2,900 m). Unlike most small rodents, this species is diurnal and is most active in the morning. Searches for insects on runways that pass under logs and through ground cover. Occasionally climbs into shrubs to feed on nectar. Adult stands on hind legs and emits an insectlike trilling for up to 10 seconds. This call, which increases in volume and ends abruptly, is probably used to define territories.

**Dusky Rice Rat** (*Melanomys caliginosus*)

8.5 in (22 cm). Similar in appearance to Alston's Singing Mouse but much larger. Tail short and dark. Common to abundant; occurs on Caribbean and Pacific slopes to about 3,280 ft (1,000 m); favors humid areas and is absent from the Nicoya Peninsula and the rest of the dry N.W. Pacific. Found in brushy fields, pastures, and forest gaps. Mainly terrestrial. Generally diurnal but also active at dusk. Feeds principally on insects, actively pursuing moths, beetles, and cicadas. Also eats fruit and seeds; sometimes forages under bird feeders if they are stocked with rice or fruit.

**Family ECHIMYIDAE (Spiny Rats and Tree Rats).** An exclusively New World family, with 90 species. These large ratlike rodents are in fact more closely related to porcupines and agoutis than to true rats. There are two species in Costa Rica. Tomes' Spiny Rat has shiny, coarse fur; the Armored Rat, not described here, has prominent, thick spines and is relatively uncommon.

## Tomes' Spiny Rat (*Proechimys semispinosus*)

17 in (44 cm). One of the largest ratlike rodents in CR. Sleek, reddish-brown fur; inconspicuous spiny hairs on the rump and upper back; belly fur is white. Bicolor tail, with scant hair, is shorter than combined length of head and body. Common to abundant; occurs on Caribbean and S. Pacific at low elevations (absent from dry N.W. Pacific). Strictly nocturnal and mainly terrestrial. Sometimes seen at night sheltered quietly among buttresses of large trees; usually moves slowly and deliberately and may remain motionless when encountered at night. During the day, sleeps in burrows or shallow depressions under roots and logs. Often takes up residence in forest encampments. Eats fallen fruit.

**Family ERETHIZONTIDAE (New World Porcupines).** The majority of an estimated 16 species (5 genera) in this family occur in South America. Interestingly, New World porcupines are more closely related to agoutis and pacas than to Old World Porcupines. The Mexican Porcupine is the sole species known to occur in Costa Rica; Rothschild's Porcupine (*Coendou rothschildi*), endemic to western Panama, may venture into the Osa Peninsula.

## Mexican Porcupine (*Sphiggurus mexicanus*)

39 in (70 cm). Pale head contrasts with black body (yellowish spines, visible on head, are largely concealed by long fur on the body). Some individuals are less hairy and show spines on rump or shoulders. Nose is pink and bulbous. Prehensile tail is nearly naked. Fairly common; occurs throughout the country except in S. Pacific. Arboreal and nocturnal; sometimes located at night by dull reddish eyeshine. During the day, sleeps in a hollow tree, leaving behind a characteristic pile of musty, oval droppings. Feeds on fruit, seed, buds, and young leaves. Generally silent, but sometimes yowls and screams during breeding season. *Illustration not to scale.*

**Family DASYPROCTIDAE (Agoutis).** Agoutis superficially resemble guinea pigs, though they are larger and have longer legs. Related to pacas, they are presently assigned to a separate family. The majority of the species in this family occur in South America. There is a single species in Costa Rica.

## Central American Agouti (*Dasyprocta punctata*)

22 in (56 cm). Large. Fur finely grizzled; upperparts orange brown (individuals in Caribbean lowlands have straw-colored rumps). The long rump hairs can be raised in alarm. Common; occurs throughout the country, on both Caribbean and Pacific slopes. Favors forests, second growth, and plantations. Solitary. Diurnal and terrestrial, this is one of the more easily seen medium-sized mammals, especially in protected areas with few large predators. In some regions it suffers from overhunting, but generally it is not prized in CR due to female's conspicuous menstrual cycle. Eats fruit and nuts; sometimes found under fruiting trees, making loud chewing sounds.

**Family CUNICULIDAE (Pacas).** There are 2 species in this family. The Paca occurs in Costa Rica and many other regions of the New World; the Mountain Paca (*Cuniculus taczanowskii*) is found only in the Andes.

## Paca (*Cuniculus paca*)

39 in (70 cm). Large, stocky. Mostly reddish brown with rows of white spots on back and sides. Common in CR but uncommon in much of its range; occurs throughout the country in or near forests. Often found along streams. A popular game animal, the Paca is rare in unprotected areas. Strictly nocturnal; freezes when caught in flashlight beam, its eyes reflecting a bright-orange color. During the day, rests in burrows excavated in banks. Forms monogamous pairs but individuals travel and den alone when not breeding. Single young are left in narrow burrow during the day and called to the entrance to nurse at night. Specialized cheekbones enable adults to produce very loud, deep barks.

## Forest Rabbit (*Sylvilagus gabbii*)

14 in (35 cm). Small. Darker than Eastern Cottontail and has shorter ears. Grayish, inconspicuous tail. (The very similar Dice's Cottontail is found only in the Talamanca Mountains.) Fairly common; occurs on the Caribbean slope and S. Pacific. Favors wet forests, where it is found in forest gaps, forested roads and trails, and other edge habitats. Solitary and mainly nocturnal; sometimes seen at dusk or dawn. Under flashlight beam, eyes appear reddish, usually with only one eye visible. During the day, rests under logs or in dense cover. Female uses dry grass to build multi-chambered nest, placed above ground at the end of a runway. Gives birth to litters of 2 to 8 young. (Formerly *Sylvilagus brasiliensis*.)

## Eastern Cottontail (*Sylvilagus floridanus*)

16 in (40 cm). Medium size. Mostly buffy brown, with orange legs and nape. Moderately long ears. Tail, cottony white below, is very conspicuous as rabbit hops away. Common; occurs in N.W. Pacific (including the Nicoya Peninsula) and on the western slopes of the Tilarán Mountains. Found in a variety of habitats though absent in mature rainforests. Mainly nocturnal but also active at dawn and dusk. Feeds on green vegetation and woody plants. Female may give birth 7 times per year, with 3 to 5 young per litter. She lines her nest with her own belly fur and usually places the nest in tall grass, either on the ground or in a shallow depression.

## Order Soricomorpha (Shrews)

The shrews, in the family Soricidae, were formerly included in the order Insectivora, but based on recent genetic evidence, the shrews, moles, and solenodons are now placed in the order Soricomorpha. There are at least four species of shrew in Costa Rica; it is very difficult to distinguish one from the other in the field.

## Blackish Small-eared Shrew (*Cryptotis nigrescens*)

3.5 in (9.5 cm). Very small. Blackish, with a short tail. Long pointed snout, continuous tooth-row, and five toes on forefeet separate this shrew from a small mouse. Common; occurs throughout the country from 2,620 to 9,510 ft (800 to 2,900 m). Found in both forests and pastures. Nocturnal and diurnal. This tiny insectivore has periods of frantic activity followed by short rests. Kills a variety of invertebrate prey and has a voracious appetite. Shrews are seldom seen because they spend most of their time tunneling under leaf litter, emerging only briefly to dash across open areas. Domestic cats often kill shrews but leave them uneaten (their scent glands make them distasteful). *Illustration not to scale.*

### Order Chiroptera (Bats)

Bats are the most diverse and species-rich group of mammals in Costa Rica, with over 106 known species. This is more than double the number of bats that occur in the United States and Canada combined, and represents about 10% of all bat species in the world. The wings consist of two very thin layers of skin stretched over elongated bones of the fingers and hand. Bats are the only mammals capable of powered flight. Nocturnal in habit, they use good vision to detect distant objects and echolocation to home in on prey or objects at close range. Elaborate folds of skin around the nostrils or mouth are used to focus outgoing calls, and large ears increase sensitivity to echoes. Bats in Costa Rica are highly diverse in diet, eating insects, blood, small vertebrates, fish, nectar, fruit, or pollen. They dominate the night skies and are indispensable pollinators, seed dispersers, and insect-pest-control agents.

In the descriptions that follow, bat measurements indicate body length only; tail measurements are not included because bats usually rest with the tail coiled beneath them.

**Family EMBALLONURIDAE (Sac-winged Bats).** Sac-winged bats occur worldwide. There are 10 species in Costa Rica. Bats in this family have large eyes, unadorned noses, and a short tail that projects beyond the tail membrane when the bat is at rest. Most roost in relatively exposed situations and are more often encountered by day than other bats. Many species have a pouch or sac on the wing (or around the tail) where the male stores bodily secretions (sweat, urine, etc.) that are used to mark his harem of females.

## Northern Ghost Bat (*Diclidurus albus*)

3 in (7.5 cm). Large. White fur; bare face; yellow ears; pale-pink membranes. Long tail membrane (with gland around the tail tip). Thumbs tiny, barely visible. Apparently rare, though difficulty in finding roosts and capturing in mist nests may bias its assessment. Occurs on Caribbean and Pacific slopes, in forests at low and middle elevations. Roosts alone or in small groups, hanging by its feet under palm fronds. Sometimes feeds on insects swarming at streetlights but usually forages above the forest canopy or high over water.

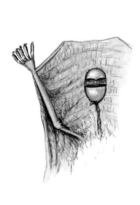
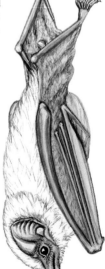

### Proboscis Bat (*Rhynchonycteris naso*)

1.75 in (4.2 cm). Very small. Grizzled gray-brown fur; two wavy pale lines on back. The only bat in CR with patches of pale fur along the forearms. Membranes are blackish brown. Common; occurs on Caribbean and Pacific slopes at low elevations. During the day, this bat is often seen along wooded streams or in mangroves. Small groups roost very close to water, often on tree trunks, the bats lined up one above the other. When disturbed by passing boats, they fly away like moths. Each group consists of a single male and his harem of females. Feeding begins before dusk; the bats fly near the water surface in search of insects emerging from an aquatic larval stage.

### Greater White-lined Bat (*Saccopteryx bilineata*)

2 in (5.2 cm). Fairly small. Blackish-brown fur; two distinct wavy lines down back. Males have saclike structures in the leading edge of the wing. Common; occurs on Caribbean and Pacific slopes at low elevations. Sometimes seen among tree buttresses or in large hollow trees. Forms groups composed of a single male and several females. Each individual within a group roosts separately in a characteristic pose, resting upside down with forearms splayed and head tilted up, clinging to the substrate with hind feet and thumbs. At dusk and dawn, the male twitters audibly and flies over his harem of females, spraying them with a strong-smelling substance consisting of urine and other bodily excretions that he stores in the wing-sacs.

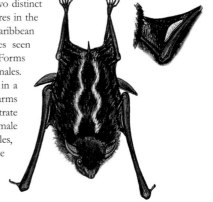

### Gray Sac-winged Bat (*Balantiopteryx plicata*)

2 in (5 cm). Small. Pale gray with dark wings; long bangs; large eyes; naked nose. Male has a small pouch in the leading edge of the wing. Common; occurs in the dry and semi-deciduous forests of Guanacaste, where it is one of the most frequently seen bats. Large groups (often 50 or more) roost near the entrance of caves or mines; in buildings; and in large hollow trees. These groups are composed mainly of males; females may roost in more secluded spots. Roosting posture is similar to that of Greater White-lined Bat. Bats leave the roost at dusk and fly above the canopy in search of insects.

**Family NOCTILIONIDAE (Fishing or Bulldog Bats).** This family has two species, the Greater Fishing Bat and the Lesser Fishing Bat, both of which occur in Costa Rica (and in most other countries in Latin America). The Greater Fishing Bat is commonly seen flying over open water; the Lesser Fishing Bat—very similar to its cousin but with smaller body and much smaller feet—favors forest streams and ponds. They are also called bulldog bats because of their split, drooping upper lips.

### Greater Fishing Bat (*Noctilio leporinus*)

4 in (10 cm). Strikingly large bat, with very long wings and huge feet. Fur orange or gray, very short; wings shiny blue-black. Easily mistaken for a night bird as it flaps wings more slowly than other bats. Uncommon; occurs on Caribbean and Pacific slopes at low elevations; seldom far from water. Has a strong, fishy odor, detectable if it passes close to observer. Uses echolocation to detect changes made by fish breaking the water surface, then catches prey by trawling with its feet. Diet also includes frogs and some invertebrates. While in flight, it transfers prey from feet to mouth, immediately consuming small animals; larger animals are roughly chopped up during flight and stored in cheek pouches for consumption when the bat comes to rest. During the day, roosts in hollow trees or sea caves.

**Family PHYLLOSTOMIDAE (Leaf-nosed Bats).** An exclusively New World family. There are about 63 species recorded in Costa Rica (more than any other bat family). Leaf-nosed bats have a triangular fold of skin around the nostrils that is thought to aid in echolocation; in a fascinating twist, they emit calls through the nose not the mouth. This highly diverse family includes carnivores, insectivores, nectar-feeders, frugivores, and bloodsuckers.

### Common Long-tongued Bat (*Glossophaga soricina*)

2 in (5 cm). Small. Fur gray brown, pale at roots. Muzzle long and narrow, with a small noseleaf at tip. Very long tongue. Tail short; extends about one-third the length of the tail membrane, which entirely encloses it. Common; occurs on Caribbean and Pacific slopes at low and middle elevations. Favors dry forests but also found in rainforests. This bat and related species are important pollinators of many forest plants. In addition to nectar and pollen, feeds on fruit and moths (and other small insects). Small to large groups roost in buildings, caves, culverts, and hollow trees; selects sites that are usually not in complete darkness. About 100 were found roosting under a bed in an abandoned cabin.

### Seba's Short-tailed Bat (*Carollia perspicillata*)

2.3 in (6 cm). Fairly small. Fur brown; each hair pale at midsection and darker at tip and root, creating a banded pattern. Muzzle moderately long. Chin with one central wart bordered by a U-shaped row of small warts. Ears and noseleaf longer and more conspicuous than in Common Long-tongued Bat. The most abundant and widespread bat in CR; occurs in gardens, clearings, plantations, and most types of forest. This bat plays a critical role in forest regeneration; it feeds on the fruit of pioneer plants such as *Cecropia* and *Piper* and then scatters their seeds in clearings as it defecates in flight. Small to large groups roost in buildings, caves, and hollow trees, and under bridges and culverts.

### Great False Vampire Bat (*Vampyrum spectrum*)

5.5 in (14 cm). The largest bat in the New World, with a wingspan of about 3 ft (1 m). Body usually orange-brown. Ears large, rounded; muzzle long with cup-shaped noseleaf; no tail; feet and claws large. Rare and local; occurs on Caribbean and Pacific slopes, usually at low elevations. Generally favors rainforests but also found in cloud forests and dry forests. This carnivore specializes on birds, particularly anis and parakeets, but also eats other bats and small rodents. Roosts in family groups, the male encircling female and young with his wings; prefers to roost in large hollow trees. Female gives birth to one young per year, usually at the start of the rainy season.

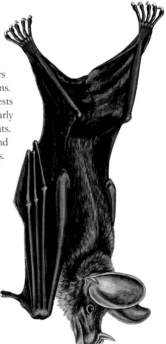

**Fringe-lipped Bat** (*Trachops cirrhosus*)

3 in (7.8 cm). Fairly large. Woolly, gray-brown fur; large, rounded ears. Medium-sized noseleaf; named for the distinctive elongated warts around mouth and on chin. Fairly common; occurs on Caribbean and Pacific slopes at low and middle elevations. Forages above streams and other bodies of water, homing in on the mating calls of certain frog species; can discriminate between the calls of poisonous and nonpoisonous frogs. Also eats insects and lizards. Small groups roost in sea caves, buildings, hollow trees, and culverts.

**Common Tent-making Bat** (*Uroderma bilobatum*)

2.5 in (6.5 cm). Medium size. Grayish fur; bold white facial stripes; pale stripe on back. U-shaped tail membrane is short and naked along edge. Common; occurs on Caribbean and Pacific slopes at low and middle elevations. Found in forests, plantations, and gardens. This species roosts in "tents" constructed out of a wide variety of palms (including coconut and fan palms) and other plants. Up to 60 bats may occupy a large tent. Eats mostly small figs and other fruits.

**Honduran White Bat** (*Ectophylla alba*)

1.75 in (4.3 cm). Very small. Fur white on upper body, pale gray on rump; yellow ears and noseleaf; blackish wing and tail membranes. Uncommon; occurs on Caribbean slope at low elevations. Found in rainforest and tall second growth. Feeds on small figs and other fruit. Small groups of 4 to 10 bats roost under "tents" made from *Heliconia* and other plants. The bats chew along each side of the midrib of the leaf, causing the sides to collapse and hang vertically, in the shape of an upside-down canoe. Small groups can be observed roosting in these tents, usually at eye level or lower. Males and females may roost together, but females with young form separate maternity colonies.

# Tent Bats

Honduran White Bats (*Ectophylla alba*).

On the Caribbean slope of Costa Rica, a walk through a lowland forest may bring you closer to Honduran White Bats than you realize. Drooping heliconia leaves are the telltale sign. Huddled beneath the leaves, these bats—resembling large balls of cotton—roost in small harems. By nipping the veins on both sides of the midrib of the leaf, they cause the sides to fold down, which results in the tent-like structure. Recent evidence reveals that both males and females participate in tent construction. The temperature inside a tent is slightly warmer and more stable than the temperature below uncut leaves, creating ideal conditions for raising young. Tents are also useful for deterring predators, as it is difficult to reach in and under a drooping leaf without causing the leaf to move and alert the bats. Squirrel monkeys, however, have learned to hunt tent bats by leaping hard onto the leaf tent and grabbing any bats that fall to the ground as the tent collapses.

In Costa Rica there are at least 10 species of bat that construct tents, using leaves from a variety of plant species to create tents with distinct architectural styles. That of the Honduran White Bat is referred to as an upside-down canoe tent.

**Jamaican Fruit-eating Bat** (*Artibeus jamaicensis*)

3 in (7.8 cm). Large, stocky. Fur gray or gray-brown; indistinct pale stripes above and below eyes. Short tail membrane and no tail. Very common; occurs on Caribbean and Pacific slopes at low and middle elevations. Found in forests, gardens, and plantations. Eats figs and other fruit, traveling 1.2 to 6.2 miles (2 to 10 km) each night between its roost and a fruiting tree. This species and other fruit-eating leaf-nosed bats are important seed dispersers that carry seeds farther from the parent tree than do birds or monkeys. Small groups roost under culverts and bridges; in caves and foliage; and in hollow trees and logs. This bat sometimes shelters under "tents" that it makes by chewing leaves of palms and banana. Also occupies tents made by other bat species.

**Common Vampire Bat** (*Desmodus rotundus*)

3.3 in (8 cm). Fairly large. Fur brown, shiny. Thickened M-shaped pad above nostrils (although a member of the leaf-nosed bat family, lacks a noseleaf). Upper incisor teeth are large and sharply pointed. Strong legs and long thumbs aid in walking and hopping on the ground as it approaches its prey. Very common; occurs throughout the country, especially where there are cattle. Feeds on vertebrate blood and appears to prefer large mammals. It does not suck but licks blood that continues to flow due to anticoagulants in the bat's saliva. Roosts in hollow trees, mines, caves, and sink holes, usually in the deepest recesses. Suitable roosts may contain up to 2,000 bats.

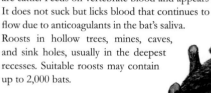

**Family THYROPTERIDAE (Disk-winged Bats).** There are four, possibly five, species in this family of tiny bats. Spix's Disk-winged Bat (*Thyroptera tricolor*) and the rare *Thyroptera discifera* occur in Costa Rica.

## Spix's Disk-winged Bat (*Thyroptera tricolor*)

1.75 in (4.2 cm). Very small and delicate. Dark-brown upperparts contrast with white belly. Suction cups under thumbs and heels. Feet diminutive, toes fused. Tail extends beyond tail membrane. Fairly common; occurs on Caribbean slope and in S. Pacific at low elevations. Seem to favor areas with large stands of *Heliconia* or shaded banana groves. Uses its suction cups to cling to the smooth surfaces of coiled young leaves; roosts facing up rather than hanging upside down as do most bats. Small family groups roost together in rolled leaves (in the form of vertical tubes) with openings of about 2 to 4 in (5 to 10 cm). These leaves are usually suitable for just one day, so the bats seek a new roost every night. Feeds on small insects captured in flight.

**Family MOLOSSIDAE (Free-tailed Bats).** This large and diverse family includes over 100 species. About 14 species occur in Costa Rica. Free-tailed bats are fast, high fliers that hawk insects. A few species occupy roofs and walls of houses and may be heard scrambling about at dusk.

## Little Mastiff Bat (*Molossus molossus*)

2.5 in (6.2 cm). Fairly small. Fur (short, velvety) is dark brown or gray-brown. Long, thick tail projects beyond tail membrane. Long narrow wings enable fast but not very agile flight. Locally common; occurs on Caribbean slope and in N.W. Pacific at low elevations (scattered records in Central Valley). Roosts in houses, hollow trees, and under palm leaves, often in groups of 300 or more. Becomes active shortly before dusk, and can be heard moving around and squeaking inside the walls or roof of a building before exiting. Flies high and fast, traveling long distances in search of beetles and other insects.

**Family FELIDAE (Cats).** There are about 40 species in the family Felidae. Six species occur in Costa Rica. Cats are strictly carnivorous, with teeth designed for tearing and slicing. They have lithe, muscular bodies. Cats in Costa Rica have 4 weight-bearing toes that have retractile claws; their tracks rarely show claw marks (unlike Coyote tracks). Though cats are seldom seen or heard, recent camera-trapping surveys indicate that several species are relatively common in some Costa Rican forests.

## Oncilla (*Leopardus tigrinus*)

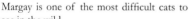

32 in (0.8 m). Only slightly larger than a house cat. Spots often form stripes on sides of body. Tail narrow, slightly longer than hind legs. Feet small, front tracks broader than hind. The few specimens of this species known from CR are all darker than the Margay. On the Margay, the "grain" of the neck fur is reversed, slanting toward the head; on the Oncilla, the neck fur slants toward the body. Rare; has been found near Cerro de la Muerte in highland oak forest and on the Osa Peninsula. Probably nocturnal. Mainly terrestrial in habit, but it is an adroit climber. Diet includes mice, shrews, and small birds.

## Margay (*Leopardus wiedii*)

39 in (1 m). Rather small and slim. Spotted (spots often form stripes on sides of body). Bushy tail is long; length exceeds length of extended hind leg. Front and hind tracks about the same size. Uncommon; occurs on Caribbean and Pacific slopes, with patchy distribution. Found mainly in undisturbed forests. More arboreal than other cats; it can partially rotate the hind feet and run headfirst down trees. Most active at night. Eats small climbing mammals such as mice, opossums, and squirrels. During the day, it rests in vine tangles (high up) or in hollow trees. Shy and elusive, the Margay is one of the most difficult cats to see in the wild.

## Ocelot (*Leopardus pardalis*)

44 in (1.1 m). Fairly large; stockier than a Margay. Spots often form stripes on sides of body. Narrow tail is shorter than hind legs. Front tracks are broader than hind tracks, giving rise to the local name *manigordo* (fat hand). Uncommon but widespread; occurs on Caribbean and Pacific slopes to high elevations. Found in dry and wet forests, second growth, and farmland with sufficient cover. Mainly active at night but may be seen at dusk or dawn. At night, travels 1.9 miles (3 km) or more in search of food. Usually hunts on the ground, though it is an adroit climber. Eats small rodents, opossums, armadillos, larger mammals on occasion; also eats reptiles, land crabs, birds, and fish. During the day, rests among buttresses of large trees and in culverts and other concealed spots.

## Jaguarundi (*Puma yagouaroundi*)

48 in (1.2 m). Long body; long, narrow tail; short legs. Unspotted; gray, blackish, or red-brown. Litter sometimes includes both gray and reddish kittens. Uncommon; occurs on Caribbean and Pacific slopes to high elevations. Found in a wide variety of habitats, including agricultural zones. In the New World tropics, this is the most easily seen cat, as it can survive in proximity to humans and is usually active by day. Sometimes raids chicken coops; also eats small vertebrates (birds, mice, and lizards) and invertebrates.

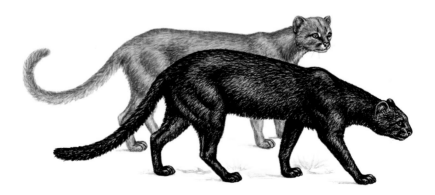

## Puma (*Puma concolor*)

66 in (1.7 m). Large. Adult unspotted; young spotted. Long legs; long tail has black tip. Broad tracks have pointed toes. Occurs on Caribbean and Pacific slopes to high elevations. Found in a variety of habitats but large tracts of forest necessary for healthy populations. Diurnal and nocturnal. Secretive, solitary, and seldom seen, but appears to be common in Talamanca Mountains and some other regions. Travels long distances, often on logging roads or trails. Unlike Jaguar, it avoids water and muddy regions. Eats medium to large mammals, including deer, paca, and agouti. Covers uneaten parts of prey with sticks and other debris, and may mark kills with urine. Usually silent; females whistle to their young and yowl when in heat.

## Jaguar (*Panthera onca*)

72 in (1.9 m). Very large; stocky; big head. Spots are in the form of discrete rosettes. Broad tracks have rounded toes. Generally uncommon to rare; occurs on Caribbean and Pacific slopes. Most numerous on the Osa Peninsula. Travels alone or in pairs. Most active at night. A skillful climber and swimmer. Eats mammals, birds, reptiles, and fish; takes small and large prey alike. Incredibly powerful, it can drag captured prey that weighs three to four times its own weight for 0.6 mile (1 km) or more.

**Family CANIDAE (Foxes, Dogs, and Coyotes).** Only two canids occur in Costa Rica, the Gray Fox and the Coyote. Members of this family lack powerful jaws and are only able to bring down large prey when hunting in packs. Most species are omnivorous. Canids leave 4-toed tracks that usually show clear claw marks (not always in the case of the Gray Fox).

## Gray Fox (*Urocyon cinereoargenteus*)

36 in (0.9 m). Small and slender. Mostly gray with rufous on legs, ears, and lower sides. Long bushy tail; tail tip and upper tail surface are black. Common; occurs on Pacific slope at all elevations; on Caribbean slope absent at low elevations. Found in a variety of habitats. Can become habituated to people and frequents some picnic areas (e.g., in Monteverde). Although nearly always seen on the ground, it is an agile climber and sometimes feeds or dens high in trees. Diurnal and nocturnal. Usually travels alone, often trotting long distances on dirt roads or tracks; pairs or small family groups are encountered on occasion. Eats arthropods, small vertebrates, and fruit. *Illustration not to scale.*

## Coyote (*Canis latrans*)

51 in (1.3 m). Large; long legs. Mostly yellow-gray; black tail tip. Uncommon to common; occurs in the Central Valley and throughout the Pacific slope (except on the Osa Peninsula). Favors agricultural areas, grassland, dry forest, and mixed-habitat areas. This species shuns mature evergreen forests and has thus benefited from human habitat-disturbance. Usually active at dawn or dusk. Can run at 40 mi/hr (65 km/hr), often traveling long distances on dirt roads and trails in search of food. Eats carrion, small mammals, sea turtle eggs, lizards, and fruit. At night (particularly in the rainy season), makes eerie, wolf-like sounds: high-pitched yips precede a long howl, the series terminating with a string of yaps. *Illustration not to scale.*

**Family MEPHITIDAE (Skunks).** Formerly considered members of the family Mustelidae, skunks are now generally placed in their own family. There are three species in Costa Rica.

### Southern Spotted Skunk (*Spilogale angustifrons*)

15 in (38 cm). Smallest of the three skunks in CR. Black with a complex and attractive pattern of white stripes and spots. Uncommon; occurs in the Central Valley and on the Pacific slope, to middle elevations. Occupies open woodlands, farmland, dry rocky terrain, and brush. Nocturnal. Emerges late at night and stays well concealed. Dens under rocks, in hollow logs, and under buildings, where its strong odor may reveal its presence; also uses burrows dug by other mammals. Mainly solitary; groups sometimes share a den. Unlike most skunks, this species can climb trees and sometimes eats fruit. Omnivorous but prefers insects and insect larvae.

### Striped Hog-nosed Skunk (*Conepatus semistriatus*)

24 in (62 cm). Fairly large. Bare pink snout; two white stripes on the back; tail black at base then pure white (shorter than head and body). Uncommon to common (rare in dry N.W. Pacific); occurs throughout the country at all elevations. Found in forest gaps, tree falls, and pastures bordering evergreen forests. During the day, lives in extensive burrows it digs at the base of trees. Solitary. Walks with tail held straight up, like a bottlebrush. Eats mostly invertebrates, some small vertebrates, and fruit.

### Hooded Skunk (*Mephitis macroura*)

20 in (51 cm). Fairly small. Pattern variable; in CR usually has a single white stripe on each side; can also have three white stripes or a single wide stripe on back. Tail (longer than head and body) is sometimes all black, very seldom pure white. Common and often seen; occurs in Guanacaste and on the Nicoya Peninsula. During the day it dens in rock crevices or burrows. Becomes active soon after dusk, wandering slowly through weedy fields and along roadsides in search of beetles and other arthropods, fruit, and bird eggs. Sometimes found at garbage dumps. Usually solitary, though several may feed together without aggression.

**Family MUSTELIDAE (Weasels and relatives).** This diverse family includes about 60 species in 22 genera. There are 4 species in Costa Rica. These carnivores typically have long bodies and short legs. Most eat meat exclusively, but some tropical species also consume fruit. Five-toed tracks show conspicuous claw marks.

## Greater Grison (*Galictis vittata*)

26 in (65 cm). Sturdy, long body; short legs. Gray above, black below; white band across the forehead and down the neck. Rare, occurs patchily on Caribbean and Pacific slopes at middle and low elevations. Usually seen in tall evergreen forests, often near streams or wetlands, but also occurs in dry, grassy regions. Diurnal and nocturnal. Terrestrial; a skilled swimmer. Dens in burrows, including those dug by armadillos. Lone animals or pairs hunt reptiles, birds, and small mammals. The habits and distribution of this elusive animal are not well known.

## Long-tailed Weasel (*Mustela frenata*)

18 in (45 cm). Very small, slinky. Fur reddish brown; white facial markings; tail has black tip. Occurs on Caribbean and Pacific slopes. Rare at low elevations; fairly common at middle and high elevations. Active by day, this weasel is occasionally glimpsed darting across a path. Long narrow body enables it to enter mouse burrows in pursuit of prey. Also hunts larger mammals such as rabbits and gophers. During the night it sleeps under rocks or in burrows made by other mammals. Lines nest with fur from prey. Individuals are usually solitary but not territorial. *Illustration not to scale.*

**Tayra** (*Eira barbara*)

44 in (1.1 m). Large. Has longer legs than most weasels and weasel relatives. Mostly blackish brown; head slightly paler than body; white patch on throat. Fairly common; occurs on Caribbean and Pacific slopes to high elevations. Found in dry forests, rainforests, second growth, and plantations. Diurnal. At night sleeps in hollow trees and in burrows. Solitary or in family groups. Travels on the ground with a bouncy gait, back and tail arched. This agile climber moves swiftly through trees, where it feeds on invertebrates, small vertebrates, and fruit. Generally silent; individuals sometimes snort in alarm, groups yowl or snarl.

**Neotropical River Otter** (*Lontra longicaudis*)

43 in (1.1 m). Long streamlined body, short limbs, thick tail, and webbed feet are all adaptations for life in water. Unlikely to be mistaken for any other mammal in the region. Uncommon to rare (endangered throughout its range); occurs on Caribbean and Pacific slopes. Found in relatively undisturbed habitats, near rivers, streams, and lagoons. Diurnal and nocturnal. Sometimes seen in Tortuguero National Park, the Osa Peninsula, and on the Corobici River. Individuals or small family groups hunt fish, mollusks, crabs, and crayfish. Deposits droppings (containing crab and crayfish exoskeletons) on rocks. Its broad, webbed tracks are sometimes encountered on sandy banks. A fast and graceful swimmer; moves awkwardly on land.

# Camera Trapping

Tayra (*Eira barbara*).

Many of the carnivores in Costa Rica, including the Tayra, are poorly known. Most are either nocturnal or arboreal, and are thus seldom encountered. One method that biologists use to study such animals is camera trapping. A pair of motion-sensitive cameras is set across a pathway, and animals that cross between the cameras trigger a photograph. This means that a variety of species can be studied without inducing the stress caused by capture.

Camera trapping studies in Costa Rica have revealed some surprises.

Pumas, often considered very rare and local, have been found to be fairly common in the Talamanca mountains and foothills. Once thought solitary, it turns out that pumas often travel in pairs. Similarly, Jaguars, although rare, are now known to be quite widespread in Costa Rica. In another instance, Northern Raccoons are strictly nocturnal in the United States, but camera trapping studies show that in Costa Rica they are most active during the day, with peak activity at 2 p.m.

**Family PROCYONIDAE (Raccoons and relatives).** This family comprises 6 genera and about 15 species, all restricted to the New World. Procyonids are omnivorous. Most species are proficient climbers. Five-toed tracks show conspicuous palm prints. There are 6 species in Costa Rica.

## Cacomistle (*Bassariscus sumichrasti*)

36 in (0.9 m). Catlike. Erect, triangular ears; long, bushy, banded tail; face black with prominent pale eye rings. Uncommon to rare; occurs in high-elevation forests, many of which are inaccessible to all but the hardiest hikers. Known from Braulio Carrillo National Park and Cerro de la Muerte. Arboreal and nocturnal. Solitary or in pairs. Travels through the forest canopy in search of fruit, nectar, insects, and small vertebrates. Loud wails (*ooyoo-whaaa* or *boyo-baa-wow*) probably used to maintain territorial spacing.

## White-nosed Coati (*Nasua narica*)

42 in (1.1 m). Combination of long muzzle (with white marks) and long, banded tail is diagnostic. Common where not hunted; occurs on Caribbean and Pacific slopes to high elevations. Found in a variety of habitats. Active by day—and highly social—this is one of the more easily seen medium-sized mammals. Erect tails waving slowly above ground cover are a sure sign that coatis are present. Lives in stable groups of 10 to 40 females and young. Males are usually solitary except when breeding. Omnivorous. Uses strong claws and probing snout to root in leaf litter for invertebrates and other food. If disturbed, an individual emits an alarm bark that causes the entire group to rush to the safety of nearby trees. These agile climbers usually sleep on branches high in the canopy.

## Crab-eating Raccoon (*Procyon cancrivorus*)

36 in (0.9 m). Orange-brown; black mask (not as extensive as on Northern Raccoon); short, banded tail. Black legs and feet. Hair on neck grows forward, toward the head. Uncommon; occurs on the Pacific slope from Quepos southward, at low elevations. Found mainly in mangroves (and other coastal areas) and swamps. Nocturnal. Rests in hollow trees during the day. Eats crabs, crayfish, snails, and fish. Usually solitary but sometimes travels in pairs or family groups.

## Northern Raccoon (*Procyon lotor*)

36 in (0.9 m). Dark brown; extensive black mask; short, banded tail. Pale legs and feet. Long white guard hairs on belly. Hair on neck grows backward. Humped back especially noticeable when it moves. Common; occurs on Caribbean and Pacific slopes to high elevations (vast range extends from Canada to Panama). Widespread in CR; found in most habitats but usually absent from extensive, mature rainforests. Solitary and mainly nocturnal; camera trapping in the Talamanca Mountains showed the Northern Raccoons there to be most active at 2 p.m. Usually travels on the ground but is a proficient climber. Omnivorous; eats whatever is easiest to catch, raiding garbage cans, corn crops, and chicken coops.

# Hanging Around

Kinkajou (*Potos flavus*).

The Kinkajou is abundant and widespread in Costa Rica. Its success is due in part to the combination of an omnivorous diet and a prehensile tail. A long tail that can grasp branches is a handy tool: it extends the Kinkajou's reach and frees up its hands for snatching fruit and small prey. Of course, a prehensile tail also decreases the likelihood that the Kinkajou will fall, an important adaptation for any animal that lives in the treetops. Of the creatures found in Costa Rica, quite a few have prehensile tails, including some of the primates; most of the opossums, anteaters, and porcupines; and a few of the small rodents. Related species found in the Old World very seldom have prehensile tails, as they principally evolved on grassy plains rather than in the forest canopy.

**Kinkajou** (*Potos flavus*)

42 in (1.1 m). Golden or gray-brown. Prehensile tail is long, tapered, and dark brown at tip. Ears, short and rounded, are positioned low on sides of head. Common; occurs on Caribbean and Pacific slopes to middle elevations. Found in dry and evergreen forests, second growth, scrub, and gardens bordered by trees. Nocturnal. At night this is the most commonly seen arboreal mammal in CR. Eyes reflect a bright orange in flashlight beam. Omnivorous; eats fruit, insects, and some small vertebrates. Usually solitary, but several may gather in a fruiting tree. Makes a variety of calls, including screams and moans; when alarmed, emits a fast, yapping bark.

**Bushy-tailed Olingo** (*Bassaricyon gabbii*)

32 in (0.8 m). Gray-brown or golden. Tail long, slightly bushy, with very faint bands. Muzzle short and pointed. (Similar to Kinkajou but smaller and lacks a prehensile tail; also more active.) Common; occurs on Caribbean slope (absent from the Pacific lowlands). Favors middle-elevation cloud forests. Strictly nocturnal and arboreal. When caught in a flashlight beam, it moves off quickly, running from branch to branch with great agility. Eats fruit, nectar, insects, and small vertebrates; sometimes feeds in same tree as a Kinkajou. Calls include a low-pitched alarm bark: *whey-chuck, whey-chuck*.

## West Indian Manatee *(Trichechus manatus)*

11 ft (3.5 m). Large, heavy aquatic mammal; can reach 1,100 lbs (500 kg). Bulbous muzzle covered in short thick bristles; rounded tail. Large flippers bear three nails. Rare; occurs along the Caribbean coast and in coastal rivers and lagoons. Endangered throughout its range; habitat loss, hunting, and fatal collisions with motor boats have all contributed to decline. Often only the nostrils are visible as it surfaces to breathe. Sometimes rests in deep water with warm undercurrents. This placid vegetarian feeds on seagrasses, water hyacinth, and other aquatic plants. Travels alone or in small groups, cruising slowly below the water surface.

**Order Perissodactyla (Tapirs)**

The four species of tapir all belong to the family Tapiridae. Baird's Tapir occurs from Mexico to Ecuador; the Brazilian Tapir and the Mountain Tapir occur in South America; the Malayan Tapir is native to Asia. Baird's Tapir is the largest land mammal in Costa Rica and the only wild representative of the odd-toed ungulates, a group that includes rhinos and horses.

## Baird's Tapir (*Tapirus bairdii*)

82 in (2.1 m). Very large and stocky; trunklike nose; short tail. Young are striped and spotted. Large tracks show three (sometimes four) triangular toe prints. Rare and endangered; occurs patchily on Caribbean and Pacific slopes to high elevations. Found in areas of extensive forest and low human populations. Commonly seen on the Osa Peninsula; also occurs in Braulio Carrillo and Santa Rosa national parks. Diurnal and nocturnal. Browses on shrubs and low trees, often feeding along streams or rivers. Also eats secondary vegetation and can benefit from selective logging in areas where hunting is controlled. Rests during the heat of the day either in mud wallows or standing water. Generally travels alone, although young may remain with the mother long after weaning. Tapirs communicate to each other with long whistles.

## Order Artiodactyla (Peccaries and Deer)

Artiodactyls, or even-toed ungulates, are significantly more diverse in Africa and Asia than in the Americas. Just two species of peccary and two species of deer occur in Costa Rica. Interestingly, whales are thought to have evolved from primitive artiodactyls, and both groups share a unique ankle bone.

**Family TAYASSUIDAE (Peccaries).** Though superficially piglike, peccaries are not closely related to domestic pigs. A large scent gland on the rump and tusks that point downward (not upward) are two features that distinguish peccaries from the "true pigs."

### Collared Peccary (*Pecari tajacu*)

36 in (90 cm). Small. Stocky body; large head; slim legs. Pale collar and dark jaw distinguish it from White-lipped Peccary. Young is paler than adult. Common where not hunted; occurs on Caribbean and Pacific slopes to high elevations. Found in forests, grasslands, and farmlands. During the heat of the day, rests in caves and large burrows and under rocks. Most activity occurs early morning, late afternoon, and at night. Its dust baths and mud wallows become permeated with characteristic rancid-cheese odor. Herds of up to 50 individuals divide into smaller groups (of 2 to 5) to feed. Eats palm nuts and other fruits, roots, and seeds; occasionally eats invertebrates. Moves quietly; makes a series of sharp woofs when alarmed.

### White-lipped Peccary (*Tayassu pecari*)

48 in (120 cm). Large. Dark coloration, with a white patch on lower lip and chin. Rare and local; confined to a few areas with extensive wilderness, including the Osa Peninsula and Tortuguero and Braulio Carrillo national parks. Diurnal and nocturnal. Usually rests at midday. Lives in large groups of sometimes more than 200 peccaries. When foraging, individuals disperse over a wide swath of forest, bulldozing through the leaf litter for roots, seeds, and fruit. Strong interlocking jaws allow it to crack open very hard palm nuts. Considered to be the most dangerous mammal in Central America, a large adult vigorously defends its young. Clicking sounds made by snapping of canine teeth are used as a warning.

**Family CERVIDAE (Deer).** Unlike sheep or goats, deer grow new antlers every year, shedding them after the mating season. In most species, including the two found in Costa Rica, only the male sports antlers. The Red Brocket, a forest species, has small unbranched antlers and a low, rounded back, two features that enable it move easily through dense vegetation. In White-tailed Deer, an open-country species, males carry large branched antlers.

## Red Brocket Deer (*Mazama americana*)

42 in (1.1 m). Rather small. Reddish brown; rounded back; tail short, whitish below. Male has short, unbranched antlers; young is spotted. Fairly common where not hunted; occurs on Caribbean and Pacific slopes to high elevations (absent from Central Valley and dry N.W. Pacific). Favors evergreen forests at low and middle elevations. Solitary. Mainly nocturnal but sometimes seen during the day. With head held low to avoid branches, moves easily through dense brush. In small clearings within the forest, feeds on fruit, flowers, fungi, and vegetation. If alarmed, remains motionless; slips away furtively; or sometimes snorts and then runs off, with short tail held erect.

## White-tailed Deer (*Odocoileus virginianus*)

48 in (1.2 m). Medium size (smaller than most races in the United States). Fur mostly gray brown to dull orange. Slender body; long legs; flat back; tail long, broad, white below. Male has large, branched antlers; young is spotted. Common; occurs on Caribbean and Pacific slopes. Favors deciduous forests and open habitats of N.W. Pacific; rare in mature rainforests. Mainly active at night or as dawn approaches. Ventures into open areas to feed on leaves and grasses; eats fruit when available. Individuals may congregate where food is abundant. If alarmed, gives a sharply exhaled snort and leaps away, the tail held up like a white flag.

birds

**With nearly 900 bird** species recorded, Costa Rica is a bird watcher's paradise. The country's small size allows visitors to travel quickly between strikingly different bird communities. An experienced birder with sufficient gumption can see 300 species in a single day, and novice birders too, with little effort, will readily see many birds.

This chapter includes descriptions of 226 species. The majority were selected because they are conspicuous or common (or both), but a few not-so-common species were also included simply because they are in some way unique. Visitors to Costa Rica may be getting their first look at tropical birds such as tinamous, guans, parrots, motmots, toucans, trogons, ovenbirds, antbirds, cotingas, and manakins. They will also encounter birds common to temperate zones—among them hawks, woodpeckers, wrens, and hummingbirds—but represented here by more species.

The trick to spotting lots of birds is to visit different areas. There are four major zones of avian diversity in Costa Rica: 1) Caribbean slope, 2) south Pacific slope, 3) north Pacific slope, and 4) highlands. Within each zone, hiking up or down a slope, just 200 yards or so, is a good way to see a new mix of birds; it also pays to explore different habitats (bird activity is often especially intense at the point where two habitats meet). Birds are much more active during the first three or four hours of the morning than during the heat of the day, so get up—and out—early. Bird activity also increases somewhat in the late afternoon, and after it stops raining.

Seeing a bird is one thing, of course, and identifying it is an entirely different matter. Binoculars are a must; any unit in the 7x to 10x magnification range is suitable. Becoming familiar with avian songs is an indispensable aid in identifying birds, especially when surrounded by dense understory vegetation. Explore the website www.xeno-canto.org to hear recordings of the songs (and calls) for most of the birds that occur in Costa Rica. Finally, even veteran birders should consider hiring a guide; a number of excellent, experienced birding guides work in the country and they will enable you to see and identify a greater number of species, in a shorter period of time.

Violet Sabrewing (*Campylopterus hemileucurus*). The largest hummingbird in Costa Rica, the Violet Sabrewing is often seen at feeders in the foothills and highlands. It occasionally chases smaller hummingbirds.

**Family TINAMIDAE (Tinamous).** Among the most ancient of avian families. Ground dwelling birds with plump bodies; almost tailless, and with thin neck, small head. Five species in Costa Rica, all resident.

## Great Tinamou (*Tinamus major*)

17 in (43 cm). Chicken size; largest tinamou in CR. Body mainly grayish brown, with fine, dark barring; bill and legs grayish. Found in mature forest at low and middle elevations of Caribbean and S. Pacific. Common (and tame) where protected from hunting, but wary elsewhere. Easily extirpated. Often solitary; searches ground for fallen fruit and seeds as well as small vertebrates and arthropods. Turquoise-blue eggs are usually laid between root buttresses at the base of a large tree. Song is a series of powerful, drawn out, tremulous whistles that gradually increase in volume; may be heard at any time of day or night (one of the most captivating sounds of the forest). Startled birds suddenly explode into flight with a rush of wings and excited whistles; the rocketing birds often crash through vegetation.

## Thicket Tinamou (*Crypturellus cinnamomeus*)

11 in (28 cm). Medium size. Legs bright red; prominent dark barring on wings, tail, and flanks. Males grayer; females more cinnamon and show bolder, more extensive barring. Common resident throughout dry N.W. Pacific and on Nicoya Peninsula; the only tinamou found in dry and scrubby forests (some overlap with Little and Great Tinamou in wetter forests of the Guanacaste Mountain Range). Usually found alone; seldom flies when startled, preferring instead to run away—or simply freeze. Call is a single, strong, clear whistle; somewhat hollow, but less quavering than the calls of other tinamous.

---

### Order Anseriformes (Ducks and relatives)

**Family ANATIDAE (Ducks).** Costa Rica has 4 resident species and 10 migrant species (most of them rare). Ducks in this country are over-hunted, persecuted as agricultural pests, and further threatened by loss of aquatic habitat.

## Blue-winged Teal (*Anas discors*)

female

15 in (38 cm). Small. In fall and early winter, plumage is brownish with pale scaling. In Jan. or Feb., adult male molts into breeding plumage; has dark bluish head with white crescent on face; note bold black-and-white patches toward rump. In flight, all birds show baby-blue wing patch. Flocks are compact, fast flying, and erratic. This is the most common migrant duck in CR. Present from late Sept. to April. Usually occurs in small flocks, except in Tempisque and Río Frío areas, where more numerous. Prefers freshwater marshes and other bodies of still water. *Illustration not to scale.*

female

male

## Black-bellied Whistling-Duck (*Dendrocygna autumnalis*)

21 in (53 cm). Neck, back, upper breast are rufous; belly, rump, and tail are black; bill, legs (long), and feet are coral red. In flight shows bold white stripe on upper wing. Sexes alike; young birds similar to adults but duller, with grayish bill and legs. Of resident ducks in CR, this is the most numerous and widespread. Found throughout lowlands in wet pastures, marshes, rivers, estuaries, lakes, and oxidation ponds. Flocks of thousands congregate in Tempisque River basin, where birds rest in marshes during the day (and at night disperse into surrounding rice fields to feed). Eats mostly seeds, shoots, aquatic vegetation, and some aquatic invertebrates. Considered a pest by rice farmers. Nests either in tree cavities or on ground amid vegetation, sometimes far from water. Costa Rican name is *piche*, or *pijije*, the latter in imitation of its strident, whistled call.

## Muscovy Duck (*Cairina moschata*)

Male 34 in (86 cm); female 25 in (63 cm). Heavy build; with broad wings and short neck. Body plumage is black with green iridescence; note white wing patch—large in males, small and indistinct in females and young. Males have red carbuncles at base of bill. Widespread in wetlands of both slopes (but often hunted); common only in lower Tempisque River basin. This arboreal duck prefers the wooded margins of wetlands (swamps, marshes, lakes, slow-flowing rivers, and mangroves). Usually seen in flight or perched in trees, alone or in small groups. Eats mostly grains; nests in large tree cavities. Free flying, domestic individuals usually show some white body feathers, have extensive reddish bare skin on the face, and are tame (wild birds are notably wary).

**Family CRACIDAE (Guans, Curassows, and Chachalacas).** Large turkeylike birds with long necks, tails, and legs; found only in tropical and subtropical regions of the New World. Sexes are alike except in curassows. Five species are resident in Costa Rica. Very sensitive to habitat loss and hunting pressure; only chachalacas regularly occur outside of protected areas.

## Gray-headed Chachalaca (*Ortalis cinereiceps*)

20 in (51 cm). Olive-brown with gray head and neck; tail dark with a broad, buff tip; throat is red and bare. Uncommon in lowlands and middle elevations of Caribbean and S. Pacific. Replaced by similar Plain Chachalaca on the Nicoya Peninsula and in parts of Guanacaste. Found in interrupted woodlands, tall second growth, and at forest edges. When traveling—in groups of a dozen or more—flocks often ascend to a treetop and then, one by one, launch into long glides on unsteady wings, in a single file. Eats fruit and tender leaves. Calls are loud and raucous; when an entire flock vocalizes, the cacophony can be heard from a great distance. In some parts of Latin America, a person who talks too much or too loudly is called a *chachalaca*.

## Black Guan (*Chamaepetes unicolor*)

25 in (64 cm). Sleek black plumage; bright blue facial skin, red iris, and coral-red legs. Endemic to CR and western Panama. Common in dense, highland forests and—when not hunted—in semi-open forests. Moves with agility through canopy and on ground; usually alone, less often in pairs, although numerous birds may congregate at fruiting trees. Eats large fruits, including wild avocados and fruits from palms, which it plucks from trees or takes from the ground. Becomes quite confiding when not threatened by hunters. Male produces a bizarre, crackling buzz by vibrating the tips of modified wing feathers during steeply descending glides through the canopy. Otherwise, not a very vocal bird (except in defense of young or nest).

## Crested Guan (*Penelope purpurascens*)

34 in (86 cm). Turkey size. Dark, olive-brown plumage shows fine white spots and is glossed with bronze and green; has bushy crest and a bare, red dewlap. Uncommon; restricted to protected forests, in lowlands and at middle elevations throughout the country. Walks and hops along forest canopy branches in search of fruit; or, less often, walks on ground and eats fallen fruit and seeds. Occurs in pairs or small family groups. In the morning and at dusk, displaying males make a deep, drumming noise with their wings as they glide from treetops. Vocalizations are loud trumpeting notes that seem to express a bird's level of stress; they become increasingly intense until the bird finally takes flight.

## Great Curassow (*Crax rubra*)

36 in (91 cm). Large and regal. Male black with white belly; base of the bill and bulbous knob on top of the bill are both bright yellow. Female rufous with fine black-and-white barring. Both sexes have tufted erectile crests. Formerly found countrywide, from lowlands to middle elevations, but now restricted to mature forests in larger parks and protected reserves. Curassows walk the forest floor in search of fallen fruit and small animals. Usually seen in pairs or small groups; but adult male sometimes encountered alone. Roosts and nests in trees. Male often flies up into a tree to call; he moves from branch to branch with a combination of awkward leaps and flaps. Breeding males produce a forceful, low-frequency humming that is difficult to locate; also produce a soft, high-pitched, descending whistle. Both sexes utter peeping notes. Adult females fiercely defend their young with pounding wings and clawing feet. *Illustration not to scale.*

female

male

**Family ODONTOPHORIDAE (New World Quails).** Of the seven species in Costa Rica, only the Crested Bobwhite prefers open areas, where it is common. All the others are primarily forest dwellers. Difficult to observe because of furtive habits and cryptically patterned plumage; listening for their loud calls is often best means of detection.

## Spotted Wood-Quail (*Odontophorus guttatus*)

10 in (25 cm). Plump bodied; mottled brown with white spots below; has a black throat and a rufous, erectile crest. Common resident in the highlands of the Central and Talamanca mountain ranges. At lower elevations, it overlaps with other wood-quail species (but it is the sole species to inhabit high mountains). In dense undergrowth, small coveys range over the forest floor, sometimes venturing into adjacent second growth or plantation fields; scratches and pecks in leaf litter for seeds, fruits, and small invertebrates. When encountered at close range, birds may freeze or scurry away one after the other, chirping nervously as they lift and lower their crests. Coveys sing a sustained chorus of raucous, whistled calls that are most often heard at dawn.

**Family SULIDAE (Boobies).** Five species recorded in Costa Rica; three of them breed here. Most species are strictly pelagic, only one is regularly seen from mainland shores.

## Brown Booby (*Sula leucogaster*)

27 in (70 cm). From shore, the one booby you are likely to see. Sleek and streamlined bird with long, pointed wings, wedge-shaped tail, long but stout neck, and conical bill. Dark brown above; dark hood contrasts sharply with white belly and underwing. Bill pale yellow (feet a brighter yellow). Males of the Pacific race have frosty-gray heads, otherwise sexes are alike. Young are dull brown, with grayish bill and dull-yellow feet. Found off both coasts although much more common on Pacific. Flies low over water—usually swiftly—with steady flaps and short glides; also flies high, in compact lines or V-formations. Follows fishing vessels. Sets down on water but more often on flotsam, including the backs of dozing sea turtles. From a few yards above the water, makes shallow-angle dives to catch fish and squid. An island nester, on bare ground or among accumulated debris; female produces a single egg. In breeding colonies, this bird's fearless disregard of humans gave rise to the common name booby.

adult

young

male Pacific race

adult

**Family FREGATIDAE (Frigatebirds).** Two species occur in Costa Rica, both of them breeding residents. (The Great Frigatebird is restricted to distant Cocos Island.)

## Magnificent Frigatebird (*Fregata magnificens*)

36 in (91 cm). Large, soaring marine bird. Long, narrow wings; deeply forked tail. Adult male black with inflatable red throat pouch (usually concealed). Adult female brownish black with white patch on breast. Young birds usually have white head and breast. Common along both coasts; abundant in Gulf of Nicoya. A graceful aerial predator, it snatches food from the water's surface or from land; also engages in acrobatic pursuit of other birds until these are harassed into giving up prey. Eats fish, squid, sea snakes, baby sea turtles, and eggs and nestlings of other seabirds. It cannot walk on land; nor can it take flight from either flat ground or surface of water (it sometimes comes to rest on trees, rocks, and the rigging of ships). Supremely adapted to life on the wing, the bird's dried skeleton weighs less than do all its feathers. Breeds on offshore islands in colonies, constructing stick nests in vegetation. Breeding male inflates his red balloonlike throat pouch during courtship display. *Illustration not to scale.*

female

young

male

male

Family **PHALACROCORACIDAE (Cormorants).** Only one cormorant occurs in Costa Rica.

## Neotropic Cormorant (*Phalacrocorax brasilianus*)

26 in (66 cm). Mostly blackish brown. Hooked bill and bulkier head and neck distinguish it from Anhinga. Takes flight from water with considerable effort, struggling over surface until completely airborne. Flaps steadily, with head held high and its neck outstretched but slightly crooked. As plumage is not waterproof, often perches to dry out spread wings. Widespread but uncommon resident along both coasts and in most lakes, rivers, and estuaries. Common to abundant resident at Lake Arenal, Caño Negro, and Tempisque River basin. Gregarious; groups sometimes float shoulder-to-shoulder, herding fish and then submerging in unison to catch prey. A colonial breeder; places stick nests on high perches in tall trees. Labeled a pest at fish farms, where cormorants are adept at foiling passive controls such as nets over ponds.

young

adult

Family **ANHINGIDAE (Anhingas).** There is a single species in the Americas; three species occur in the Old World.

## Anhinga (*Anhinga anhinga*)

34 in (86 cm). Long thin neck (with distinctive crook), small head, needlelike bill, long tail. Adult male is black with silvery white spots and streaks on its wings and back; tail dark with a wide, buff tip. Female has a blackish body that contrasts with buff-brown head, neck, and breast. Young are similar to female but browner overall. Flies with head fully outstretched; when soaring high overhead, looks like a flying cross. Widespread but uncommon resident in lowlands; occurs in forested margins of lakes, lagoons, large rivers, and estuaries. Most numerous around Caño Negro and Tempisque River basin. Seen singly or in pairs. Uses bill to spear fish underwater. Often swims with body submerged and long snakelike neck above water. Must crawl onto log to dry soaked plumage before being able to fly. Spanish name is *pato aguja* (needle duck).

male

female

**Family PELECANIDAE (Pelicans).** Two species occur in Costa Rica; one is a common breeding resident, the other an extremely rare migrant.

## Brown Pelican (*Pelecanus occidentalis*)

43 in (109 cm). Very large. Body brown with gray highlights; head and neck white. Long, brown bill and a pendant throat pouch. Note: during breeding season, crown yellow, nape chestnut, and bill pinkish. Young are brownish overall except for white belly. Common to abundant along entire Pacific coast, where it breeds in a few colonies; an uncommon, nonbreeding visitor on the Caribbean coast. Awkward at best on land, but flies with apparently effortless flaps and long glides, gracefully skimming the tops of waves; when higher up, birds generally form one side of V-formation—or, less often, a complete V. Captures fish (its principal prey) with spectacular, twisting crash-dives; folds back its outstretched wings just as the bill pierces water; bobs back to the surface, strains water from its extended pouch, and gulps down any prey. Nests on islands; places its stick nest in a tree or shrub.

nonbreeding
adult

breeding
adult

**Family ARDEIDAE (Herons and Egrets).** To date, 19 species (residents and migrants) recorded in Costa Rica. Predatory birds that spear or grasp prey with pointed bill. Fly on bowed wings, with neck folded back and legs trailing behind. Sexes are alike. When breeding, they develop ornate plumes and brightly colored bare skin.

## Bare-throated Tiger-Heron (*Tigrisoma mexicanum*)

32 in (80 cm). Long thick neck, short legs, cryptic barred plumage; holds head horizontally. All ages show bare yellow throat. Has labored wingbeats, often flies with neck outstretched. When foraging, stands still or moves very slowly. Common resident in lowlands, where it prefers mangroves, wide rivers, freshwater marshes, and wet pastures. Seldom far from tree cover. Of the three species of tiger-heron in CR, Bare-throated is seen most often. Rufescent is restricted to swamp forests of the Caribbean lowlands; Fasciated is found along swift, rocky streams (both have feathered white throats). During breeding season, gives deep, roaring "swamp monster" calls at dusk or later. Solitary breeder that nests high in tall living trees.

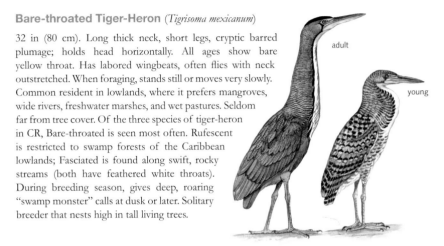

## Great Blue Heron (*Ardea herodias*)

52 in (132 cm). Largest heron in CR. Sturdy build. Gray body, whitish head, black crown plume, black lines on foreneck; rufous thighs. Slow, stiff wingbeats and frequent glides. Note two-toned upper wing in flight. Stands quietly in still water to watch for prey, then makes lightning stab. Widespread at low and middle elevations, in diverse aquatic habitats. Does not breed in CR but found here year-round. Individuals usually alone or widely separated; however, a few birds may congregate around food sources and this species is known to migrate in small flocks. Often active at night. Utters a harsh croak when startled into flight.

## Great Egret (*Ardea alba*)

40 in (101 cm). Largest all-white heron in CR. Tall and slender. Pure white plumage at all ages. Yellow bill and black legs. When foraging, waits motionlessly for prey—or stalks slowly. Maintains an erect posture. May wade into deep water with legs almost completely submerged. Great Egret is found countrywide, in lowlands and occasionally at higher elevations. Mostly a nonbreeding migrant from the north, but a small breeding population resides here. Inhabits a variety of freshwater and brackish habitats. Alone or in large groups, with individuals evenly spaced within the flock.

## Snowy Egret (*Egretta thula*)

24 in (61 cm). Entirely white (of the white herons in CR, this is the one most often seen on beaches). Slender build; black legs/yellow feet, thin black bill, and yellow facial skin. Young Snowy Egret is very similar to young Little Blue Heron but the former has yellow feet, some black on front of its green legs, and a trace of yellow at base of bill. A fairly active forager that often stands quietly, then spreads its wings and hops in pursuit of small fish. Widespread in lowlands; this nonbreeding migrant is a common resident in winter but less common in summer. Found alone or in groups.

## Cattle Egret (*Bubulcus ibis*)

20 in (51 cm). Small white egret with chunky build, black legs, and short yellow bill. When breeding (April to Nov.), note reddish bill and legs, and buff stains on crown, neck, and back. Common resident countrywide; restricted to lowlands when nesting but ranges to higher elevations when not breeding. Eats mostly insects, which it catches by following after cattle and tractors. Gregarious. Forages in small flocks; roosts in large concentrations, typically in trees that overhang water (huge colonies nest on wooded islands in rivers and estuaries). Some time after cattle ranching was established in South America, the Cattle Egret colonized the New World, apparently having crossed from Africa on its own. In CR, it was first recorded in 1954.

breeding

nonbreeding

# Open-country Aliens

Cattle Egrets (*Bubulcus ibis*) in full breeding plumage.

Originally from Africa, Cattle Egrets first showed up in South America in 1877, and by 1954 they had spread to Costa Rica. Today, they are firmly established throughout the Americas. With logging, cattle ranching, and other causes of deforestation continuing apace, the open spaces where these birds thrive have also grown in number. In Costa Rica, they first nested on Pájaros Island, in the Tempisque River; in the late 1970s they outgrew that initial breeding ground and began forming new colonies at other lowland sites, where today the movement of Cattle Egrets from nocturnal roosts to feeding areas is a common sight.

Other birds that are native to open-country habitats—north (or south) of Costa Rica—have also invaded the country's once forested lands. To date 16 species have established residency here, the majority having crossed over from Panama. Meanwhile, the populations of many forest birds have plummeted and some, like the Orange-breasted Falcon, are now extinct in Costa Rica.

## Little Blue Heron (*Egretta caerulea*)

24 in (61 cm). Somewhat stockier than herons of similar size. Bill gray with dark tip; legs and feet are pale greenish. Young are entirely white (compare with young Snowy Egret); adults are dark gray but note maroon tinge to neck and head. Birds of intermediate age are a distinctive, patchy gray and white. Usually alone. Moves slowly when foraging, with neck inclined forward and bill pointing down. Little Blue Heron is widespread and common. Occurs as both migrant and nonbreeding resident (though small numbers occasionally nest here). Found at low and middle elevations in a variety of aquatic habitats, but seems particularly fond of both riverbanks and shallow, freshwater bodies with emergent vegetation.

adult

first year

## Tricolored Heron (*Egretta tricolor*)

26 in (66 cm). Neck and bill are notably long and slender. Combination of dark, slate-blue coloration above and contrasting white belly is distinctive. Reddish wash on back and chest. Often forages actively, dashing and prancing after prey; at other times, stands quietly while opening its wings to shade the water—and thereby attract fish. Widespread but uncommon in lowlands (occurs as both migrant and nonbreeding resident). More numerous near the coast. Usually only one or two individuals are seen at a time, even among large congregations of other species of herons and egrets.

young

adult

## Black-crowned Night-Heron (*Nycticorax nycticorax*)

25 in (64 cm). Big head; short bill, neck, and legs. Adult has white body, black cap and back, gray wings and tail, red eyes, and yellow legs. (cf. Yellow-crowned Night-Heron.) Young is brownish and streaked primarily with white. Common resident in lowlands of N. Pacific. On Caribbean slope, widespread but much less common. Some northern migrants occur here (Oct. to March). Prefers freshwater marshes and other wetlands. Colonies nest in trees overhanging water. During day, flocks roost in reeds or dense trees; at night, solitary individuals forage at edges of still bodies of water. Eats mostly fish. At night, flying birds emit a barking *wok*.

adult

young

## Yellow-crowned Night-Heron (*Nyctanassa violacea*)

24 in (61 cm). Gray body; head is boldly patterned in black and white; orange eyes. Distinguished from Black-crowned Night-Heron by thicker bill, thinner head and neck, and longer legs. Young is brownish and primarily spotted with white. Regular resident on both coasts, in mangroves. Much less numerous inland, along big rivers. Northern migrants augment local population between Oct. and April. Forages at night in saltwater or brackish habitats; typically roosts by day in mangroves (though occasionally active). Mostly eats crabs. Nests in solitary pairs or in low-density colonies. Usual call is an accelerating series of *kwok* notes.

adult

young

**Green Heron** (*Butorides virescens*)

17 in (43 cm). A small, compact, dark heron. Gray-green body, reddish neck, and dark cap. During breeding season, yellow legs turn orangish. When agitated, raises shaggy crest and pumps short tail. Crouches and waits—or stalks slowly—to forage. Common and widespread resident at low and middle elevations; some North American migrants winter in the country. Favors wooded margins of shallow waters (ponds, rivers, and mangroves). Usually solitary. Known to carry a feather or other floating object to use as a fishing lure.

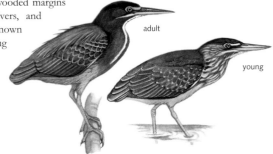

adult

young

**Boat-billed Heron** (*Cochlearius cochlearius*)

20 in (51 cm). Chunky bird with big head, large dark eyes, and bizarre, boat-like bill. Floppy black crest and white forehead; body is tawny gray with black on flanks and underwing; belly is a rusty color. (Large eyes and touch-sensitive bill are adaptations to nocturnal feeding.) Throughout lowlands, a local resident of mangroves and wooded margins of rivers. Best seen during the day, when birds rest on branches in dense vegetation at water's edge. When disturbed makes a laughing or clucking series of excited guttural notes; local name of *chocuaco* is based on vocalization.

adult

young

**Family THRESKIORNITHIDAE (Ibises and Spoonbills).** Five species recorded in Costa Rica. Gregarious waders with long neck and legs. Strong fliers that move with shallow wingbeats and outstretched neck (often glide but never soar). Flocks fly in a line or in V-formation. Sexes alike; in many species, plumage of adults differs from that of immatures.

## White Ibis (*Eudocimus albus*)

25 in (63 cm). Adult white, with black wing tips. Long, curved bill (bill and legs are red). Young is white below, brown above and on neck; when older, patchy white and brown; bill and legs pinkish. Color of bill, legs, and even plumage may be obscured by mud. Forages in groups; walks with head down, bill probing mud. Resident, found in lowlands on both coasts, but common only around Gulf of Nicoya, Caño Negro, and, locally, on S. Pacific; in soft mud of saltwater or freshwater habitats. Eats mostly small crustaceans, especially crabs. Colonies breed on small islands, in trees. Rather quiet; only calls are nasal grunts and muffled honking notes.

adult

young

## Green Ibis (*Mesembrinibis cayennensis*)

22 in (56 cm). Uniformly dark. Bulky with fairly short neck and legs. Usually appears blackish, but can show bronzy-green tones on neck and breast. (Glossy Ibis is slimmer, wings chestnut, tends to form larger flocks, and frequents open, marshy habitat; Green Ibis occurs only in forested wetlands.) Resident in forested Caribbean lowlands and foothills; generally uncommon but locally common. Usually seen alone or in pairs, occasionally in small groups; perches in tall trees, on river bars, and in swamps. Sometimes forages in open, wet fields bordering forest; pokes into mud in search of small invertebrates. Very vocal; calls in flight—at dawn and dusk—or when flushed; makes a growling, mellow yodel that is deep and far carrying.

**Roseate Spoonbill** (*Platalea ajaja*)

32 in (81 cm). Only pink bird in CR. Bare head; long straight bill with spoon-shaped tip. Plumage white (young) to pinkish (adult); tail is orange; tail coverts and bend of wing are carmine red in adult. Common resident in Tempisque River basin, Gulf of Nicoya, and, locally, elsewhere in Pacific lowlands. On Caribbean slope, common only in region of Caño Negro. Prefers shallow, open water (both fresh- and saltwater); moves location as water levels change—and young may disperse widely. When feeding, walks slowly through water; submerges its partially opened bill and sweeps head from side to side, all the while flushing out prey with feet; bill snaps shut on contact with prey. Often forages in groups. Eats fish and aquatic invertebrates. Pink color derived from diet; location and intensity of color is a function of age. In CR, a single breeding colony occurs on Pájaros Island in the Tempisque River. *Illustration not to scale.*

adult

young

**Family CICONIIDAE (Storks).** Two species in Costa Rica. Large, long-legged waders with long, stocky necks and heavy bills. Head and neck are bare. Strong fliers that may soar to great heights. Fly with neck and legs extended. Sexes alike. Adults mostly silent but nestlings are quite vocal.

## Wood Stork (*Mycteria americana*)

40 in (102 cm). Has white plumage but note black flight feathers; stout, dark bill is slightly drooped at tip; blackish head and neck are both bare. Young have light-colored bill; downy feathering on head and neck. Soaring Wood Stork resembles King Vulture (p. 64) but note stork's protruding neck and feet. Common resident in lowlands of Guanacaste and, seasonally, in Caño Negro area. Wanders widely over lowlands of both slopes. Inhabits a variety of fresh- and saltwater habitats. With open bill probes shallow water—also sweeps water. To startle prey from cover, stirs water with one foot then the other, and also partially opens a wing. Eats mostly fish, but also a wide variety of other vertebrates and invertebrates. Occurs alone or in large flocks. A colonial breeder; places its flimsy stick nest in trees that overhang water.

adult

adult

young

## Jabiru (*Jabiru mycteria*)

55 in (141 cm). Huge; hulking size dwarfs other waders. Plumage all white; black bill is massive, pointed, and slightly upturned; head and upper neck are bare and black; base of neck is red. Still breeds in Tempisque River basin, though increasingly rare there; a seasonal visitor to the Caño Negro area. Found in freshwater marshes and wet agricultural areas. When feeding, stalks slowly, then stabs or picks at prey. Is especially fond of eels but eats any other small vertebrate it can catch. Usually alone or in pairs but may congregate at food sources. Solitary breeder; pairs construct a platform nest—often far from water—in tall trees within forests. Claps bill when disturbed. In CR, it is highly endangered.

**Family CATHARTIDAE (Vultures).** Four species in Costa Rica. Carrion eaters that spend a lot of time soaring. They have bare heads, strong bills, and weak feet. Lack a syrinx and are essentially voiceless. Often perch with wings spread. In flight, occasionally flex their wings below the body, a behavior unique to vultures.

## Black Vulture (*Coragyps atratus*)

25 in (64 cm). Stocky build; plumage and bare head are black; legs gray. In flight shows pale outer primaries on underwing. Distinguished from Turkey Vulture by broader wings and shorter tail; also flies with a flatter wing profile. Soaring and gliding is interrupted by quick, choppy wing flaps. Resident; abundant throughout the country, mostly over inhabited, cleared terrain; less numerous in extensively forested areas. Soars to great heights to search visually for carrion, often keying in on actions of other vultures. Also kills defenseless prey such as hatchling sea turtles and eats oily or starchy fruits. Gregarious. Forms large flocks around trash dumps and sizeable carcasses; such congregations occasion much aggressive hissing, grunting, and hopping about. Does not build nest—lays eggs on ground, under cover of dense vegetation or on rocky ledges inaccessible to most predators.

## Turkey Vulture (*Cathartes aura*)

30 in (76 cm). Long, slim build. Blackish-brown plumage; small, red head; white bill. Young have blackish head. In flight, underwing has pale, silvery flight feathers and black wing lining. (cf. Black Vulture.) When foraging—alone or in small numbers—glides low, the wings held in a slight V-shape (but seldom flapped) as the body rocks lightly from side to side; wingbeats are slow and cumbersome. Common countrywide; resident and migrant populations. In migration, spectacular flocks form, soaring and gliding high overhead. Finds carrion with its acute sense of smell.

## King Vulture (*Sarcoramphus papa*)

32 in (81 cm). Large, bulky bird. Creamy white plumage; black flight feathers. Head is a many-colored, garish affair; bill is orange; irises are white. Young are entirely black and take four years to acquire adult plumage. Very broad wings and short tail in flight; head appears small; long primaries form prominent "fingers" at wing tips. Soars on flat wings with wing tips raised; deep, powerful wingbeats. Tends to fly apart from other vultures. Rarely seen perched. Resident; uncommon to rare countrywide, in lowlands and foothills. Requires extensive forested tracts but also ranges widely over open country. Solitary or in pairs. May take small live prey. Dominates other vultures at a carcass.

young

adult

adult

young

adult

adult

**Family ACCIPITRIDAE (Kites, Hawks, and Eagles).** Diurnal birds of prey. Thirty-eight species in Costa Rica. All have hooked bills and strong feet with sharp talons. Plumage changes with age, but sexes often similar (though females are larger than males). In larger species, immature birds pass through a complex sequence of plumages.

## Osprey (*Pandion haliaetus*)

23 in (58 cm). Body is dark brown above, white below. Mostly white head shows dark eye stripe. Light underwing bears a prominent, dark wrist mark; in flight, narrow wings are held in an M-shape. Passage migrants are widespread and fairly common. Winter residents common; summer residents rare; winter and summer residents occur along both coasts and, inland, at bodies of water (lakes, estuaries, and slow-moving rivers—where clear, still water affords successful fishing). Also feeds at fish farms. Hunts by hovering high over water and then plunging feet first to grab fish; typically eats and rests at a high exposed perch, often at water's edge. Feet have rough spiny surface, reversible outer toes, and long rounded talons: these are all adaptations for grasping slippery fish. Calls frequently, issuing a series of loud, ringing whistles: *kyew kyew kyew...*

## Swallow-tailed Kite (*Elanoides forficatus*)

23 in (58 cm). Snowy white; black wings; black, deeply forked tail. Superbly graceful and acrobatic in flight. Common breeding resident, in foothills and mountains; migrants are common throughout lowlands for much of the year (but mostly absent between Sept. and Jan.). Catches flying insects, but will also snatch prey (insects, lizards, snakes, and birds) off vegetation. Can eat small prey (grasped in one talon) as it soars leisurely. Sometimes, while flying, makes a game out of dropping and then recapturing a piece of fruit or a sprig of leaves. Nest is placed high in tall, isolated trees, often on a hilltop. Outside of the breeding season, sometimes forms large communal night roosts. Calls frequently in flight, especially when in groups—and when mobbing larger raptors.

## White-tailed Kite (*Elanus leucurus*)

16 in (41 cm). Has a light build; wings and tail are long. White plumage; gray back and wings; black shoulder patch. Young faintly streaked below; have brown scales above. Buoyant, gull-like flight; glides on raised wings. Resident; common countrywide, at low and middle elevations. Fond of open areas with scattered, tall trees: pastures, grassy fields, and agricultural land. When hunting, hovers, lifts wings over back, and then parachutes down on prey; eats mostly small rodents, and some lizards, large insects. In Costa Rica, first reported in 1958—having expanded its range from countries to the north. Has moved south into Panama, where deforestation has created ideal habitat for this species.

adult

adult

young

## Snail Kite (*Rostrhamus sociabilis*)

17 in (43 cm). In all plumages shows broad, rounded wings. Square tail is dark but has white base and white, narrow tip; long, thin, sharply hooked bill; facial skin and legs are orange to red. Male slaty gray. Female dark brown above, heavily streaked below; face bears pale line above eye. Young similar to female but paler brown. Flies low on bowed wings; slow, floppy wingbeats. Resident; uncommon in the Tempisque River basin and Caño Negro region, in open marshes and lagoons. Often perches on flimsy marsh vegetation. Changes location in response to changing water levels. Eats aquatic snails that it plucks from surface with foot and then carries in bill. A social bird, as its scientific name suggests. Birds roost and nest in low-density colonies. Call is a sputtering, nasal rattle.

adult
male

adult
male

young

## White Hawk (*Leucopternis albicollis*)

22 in (56 cm). All-white hawk, except for some black on wings and a single black band on tail. Heavy build; broad wings and rather short tail; eyes are black; bare legs are yellow. Resident; uncommon on Caribbean and S. Pacific slopes, at low and middle elevations. Prefers forested hills; perches quietly in canopy or at edge of forest clearing. Sometimes shadows troops of monkeys to catch prey that these flush out. Eats a variety of small vertebrates. Can be quite confiding. On flat wings, soars in a slow circle, with occasional lazy flaps. Soaring birds call often, emitting a drawn-out, scratchy scream.

## Common Black-Hawk (*Buteogallus anthracinus*)

22 in (56 cm). Wide wings and long legs. Adults black but note single white band on tail. Young: above brownish with lighter barring; below rufous and buff with heavy dark streaks; tail dark with fine light bars. Eyes dark, legs and base of bill are yellow. Very similar to Great Black-Hawk. Common resident of lowlands, in mangroves; riparian and swamp forests; and wooded margins of marshes, lagoons, estuaries, and beaches. Seldom far from water. Hunts most often from a low perch, sometimes on the ground or, rarely, while wading in water. Pounces on prey or runs after it. Mainly eats crabs but also reptiles, frogs, fish, and hatchling sea turtles. Generally soars on flat wings and with tail spread. Quite vocal when perched and in flight; call is a series of staccato whistles that trail off at the end. (Populations on the Pacific coast formerly treated as a separate species— the Mangrove Black-Hawk—but now considered a subspecies only.)

adult

young

adult

### Double-toothed Kite (*Harpagus bidentatus*)

13 in (33 cm). Small raptor. Grayish brown above, pale below; dark tail with three narrow white bands; note dark central streak on white throat. Underparts barred rufous in adult; streaked with brown in young. Breast of female almost entirely rufous. Soars high; with wings slightly bowed and tail unspread, white undertail feathers puffed out. Resident; common countrywide in forests of low and middle elevations (though scarcer in dry N.W. Pacific). Tends to perch inside the forest canopy. Habitually shadows monkey troops, especially White-faced Capuchins, to snatch large insects or lizards flushed out by the primates. Call is a weak two-noted whistle, the first note short, the second long.

adult

young

adult

### Roadside Hawk (*Buteo magnirostris*)

14 in (36 cm). Small raptor; with short wings but fairly long tail. Gray on breast and above; rufous and buff barring on belly; rusty wings and tail. Young similar, but browner, and with streaking on breast and barring on belly. Pale yellow eyes help distinguish this species from Broad-winged and Gray hawks. Resident; common on both slopes, in lowlands and foothills. Occurs in broken woodlands and in open agricultural areas with scattered trees; except in dry N.W. Pacific, seldom found in forests. Aptly named, as it often hunts from perches on roadside trees and utility lines. Eats a wide variety of invertebrates and small vertebrates. Very vocal; utters either a drawn-out, squealing *kreeeaa*, or a series of barking *yek* notes.

adult

adult

adult

young

## Broad-winged Hawk (*Buteo platypterus*)

15 in (38 cm). Small; looks chunky when perched. Adults are brown above; creamy below, with brown to rusty barring; breast often solidly marked. Tail with black and white barring. Young lightly streaked below; tail finely barred gray and brown. Eyes are brown to gray-brown. In flight, pale underwing shows no markings; wing tip and trailing edge are dark. Soars with wings level or lifted slightly. Occurs countrywide; winter residents (Oct. to April) are common, migrants abundant. In migration, travels in huge flocks, generally along the Caribbean coast or over the central mountains. During winter, prefers semi-open country with tall trees, though it occasionally enters forests. Hunts from perch; eats small mammals and reptiles, less often insects and birds. Its piercing whistled call (*teeyeeeee*) is often heard when wintering birds defend territory.

adult

young

adult

## Gray Hawk (*Buteo nitidus*)

17 in (42 cm). Above pearly gray; below shows fine gray and white barring (birds in S. Pacific also with barring on back); tail black with white bands. Young are brown above, streaked and spotted below; head with strong dark-and-light markings; tail brown with fine barring. Eyes medium to dark brown. Resident; common in lowlands of dry N.W. Pacific and in northern half of Caribbean slope; much less common in lowlands of S. Pacific. Favors riparian forests and semi-open country with tall trees. Often perches along roadsides. Sometimes soars but usually seen in flap-and-glide flight. Hunts from a high perch or—in flight—dashes after prey; primarily eats reptiles but also small birds and mammals, some insects. Calls include a hauntingly whistled *huuuweeo* and a more typical harsh scream.

adult

adult

young

**Ornate Hawk-Eagle** (*Spizaetus ornatus*)

25 in (63 cm). Adult with long erectile crest and striking coloration. Young have shorter crest; plumage mostly white, with varying amounts of black barring on legs and flanks. All ages have feathered legs and a long, barred tail. Immature birds sometimes mistakenly identified as the much larger and extremely rare Crested Eagle, but that species has bare legs and never soars. Resident; uncommon to rare on Caribbean slope and in S. Pacific; lowlands and foothills, and, locally, into highlands. Appears regularly only within extensive forests, although it sometimes wanders into disturbed areas. Hunts alone; perches quietly inside the forest, then makes a powerful dash for prey; feeds on large birds and mammals, sometimes reptiles. Can be quite vocal when soaring; flies freely above the forest, often in pairs.

adult

young

adult

**Family FALCONIDAE (Caracaras, Forest-Falcons, and Falcons).** A diverse family of diurnal predatory birds. Thirteen species have been reported from Costa Rica. Four species of falcon are migrants, all others residents. The Orange-breasted Falcon is now considered extinct here. Falcons of the genus *Falco* have a toothlike notch on the upper mandible that is used to sever the spinal cord of prey.

### Collared Forest-Falcon *(Micrastur semitorquatus)*

Male 20 in (51 cm); female 24 in (61 cm). A slim raptor. Legs and tail very long; wings short. Tail has bold black-and-white barring; dark upperparts interrupted by a pale collar. Adults are whitish or buff below; young show heavy, dark scalloping below. Resident; widespread but uncommon in lowlands and foothills; appears to be more common in the dry N.W. Pacific. Prefers tall forest or second growth; perches fairly low in canopy when hunting, higher up when vocalizing. Does not soar. Spanish name is *gateador* (the crawler), from its habit of entering dense vegetation on foot to capture prey; eats reptiles, birds, and mammals, and occasionally large invertebrates. More often heard than seen, its call is a shouted *owhh*, given frequently at dawn and dusk.

adult pale morph

adult dark morph

young

### Crested Caracara *(Caracara cheriway)*

24 in (61 cm). This falcon looks like a hawk and behaves like a vulture. Crest black; cheeks and neck white; shoulders barred with black; wings and body black. Stout bill is grayish, cere and bare facial skin are reddish. Mostly whitish tail has a broad, black tip. Sexes are alike. Glides with wings crooked and bowed, showing white patch at base of primaries. Resident; common in open country of dry northern lowlands, becoming less common to the south. Primarily a scavenger but occasionally takes live prey or pirates prey from other birds, especially vultures. Seen on the ground or perched in trees. Often cruises above highways in search of roadkill. Unlike other members of the falcon family, this species constructs its own stick nest, usually placed in an isolated palm or other tall tree. Vocalizations are dry rattles. Spanish name is *cargahuesos* (the bone carrier).

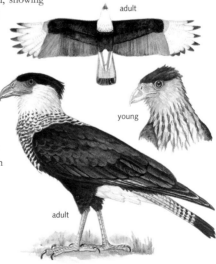

adult

young

adult

## Laughing Falcon (*Herpetotheres cachinnans*)

21 in (53 cm). Robust. Large blond head has a striking dark mask. Yellow-headed Caracara is slimmer and has only a slight mask. Resident; fairly common on both slopes at low elevations. Usually seen perched, intently scanning the ground below for snakes, its preferred prey. After nipping off the snake's head, it sometimes swallows the entire body as if it were a string of spaghetti. Apparently has some immunity to snake venom. Prefers to nest in tree hollows. Named for the peal of laughing notes that initiate its lengthy song; pairs often sing in duet fashion. Local name *guaco* refers to the sustained, two-part *gua-co* phrases of its song. Often sings well beyond sundown.

## Bat Falcon (*Falco rufigularis*)

Male 10 in (25.5 cm); female 12 in (30 cm). A small, dark falcon with long, pointed wings. Blackish above; white throat; rufous thighs and undertail. Wing flaps are shallow; in flight, it closely resembles a White-collared Swift, p. 93. Resident; uncommon on Caribbean slope and in S. Pacific, at low and middle elevations; rare in dry N.W. Pacific. Apparently requires forest nearby, though it is regularly seen in open areas, even towns. Encountered alone or in pairs; tends to perch on high, exposed branches. Catches prey on the wing, overtaking it in rapid yet short pursuit; sometimes skims along the ground at high speed in an attempt to ambush prey. Generally hunts at dawn and dusk; in addition to bats, it also eats birds (particularly swallows and swifts) and flying insects such as dragonflies. Very vocal, especially in flight and when harassing larger raptors.

**Family RALLIDAE (Crakes and Rails).** Fifteen species in Costa Rica, two of them migrants. Shy but vocal; notoriously difficult to observe. Ground-dwelling inhabitants of marsh, pasture, and forest thicket. The laterally compressed bodies of rails allows them to slip easily through dense vegetation.

## White-throated Crake (*Laterallus albigularis*)

6.5 in (15 cm). Small. Has rufous upperparts and fine black-and-white barring on flanks and undertail. Caribbean birds have a gray hood; Pacific birds have an olive crown. Resident; widespread and common on Caribbean and S. Pacific, at low and middle elevations; occurs in wet, open areas with low, dense, grassy vegetation that provides concealment: marshes, flooded pastures, and roadside ditches. Most often encountered alone, though seldom seen; occasionally seen running (or flying weakly) across a path—or as it peeks out from the edge of cover. Local name *huevo frito* (fried egg) refers to sputtering call that sounds like a fried egg sizzling in hot oil. Rarely sighted, this bird's oft-heard vocalization is the true indicator of its abundance. *Illustration not to scale.*

Pacific race

Caribbean race

## Gray-necked Wood-Rail (*Aramides cajanea*)

15 in (38 cm). Gray head and long gray neck; yellow bill; red eyes; olive back and wings; rufous breast; black belly, undertail, and tail; long, pinkish-red legs and feet. Sexes similar. Resident; common countrywide at low and middle elevations, in forests, woodlands, and agricultural lands. Prefers mangroves, swamps, and banks of streams. Generally favors dense vegetation but will appear in trails, mudflats, yards, and other open areas. Walks cautiously, constantly flicking a stubby tail as it probes soil or throws aside vegetation in search of small invertebrates, frogs, seeds, and fruit. Song is a loud, repeated *cook, cook, cook, cook-it, cook-it, cook-it...*

## Purple Gallinule (*Porphyrio martinica*)

13 in (33 cm). Unmistakable; note striking, iridescent bluish purple on most of body. Bronzy olive on back, wings, and tail; white undertail; pale-blue frontal shield; bill is red with a yellow tip. Sexes similar. Young are pale brown, with greenish wings. Widespread in lowlands and at middle elevations; generally not very abundant, but common in large marshes such as Palo Verde. Prefers habitats with emergent aquatic vegetation: freshwater marshes, wet pastures, and pond margins. Found singly or in noisy family groups. Scrambles awkwardly through thickets and tall reeds; walks on floating mats of vegetation; rarely swims. Flies weakly, teetering to the ground with outspread feet. Eats fruits and seeds, aquatic insects, frogs, fish, and, sometimes, the eggs or young of other birds. Very vocal; makes loud screeches and harsh reedy peeps, especially when startled. Young birds help parents care for hatchlings.

young

adult

**Family ARAMIDAE (Limpkin).** The lone member of this family is the Limpkin, a bird closely related to both cranes and rails. It is confined to tropical and subtropical regions of the Americas.

## Limpkin (*Aramus guarauna*)

26 in (66 cm). Heronlike, but note plump body and long, slightly decurved bill (yellow). Plumage brown with white streaks and spots. Ages and sexes similar. Flies with loose yet jerky wingbeats (and with legs dangling). Resident; uncommon in Tempisque River basin and Caño Negro region; elsewhere in lowlands it occurs sporadically. Inhabits freshwater marshes, open margins of ponds and slow rivers, and, less often, swamps. Solitary and inconspicuous when foraging. Walks with deliberate, high steps, picking and poking in search of aquatic snails. Carries snails to a low perch then pecks open the shell. Discarded shells pile up under favorite perch. Very loud, wailing cries heard mostly at dawn or dusk. Looks like a heron or ibis, behaves like a rail, but it is most closely related to cranes.

**Family BURHINIDAE (Thick-knees).** This mostly Old World family includes just nine species. There are two species in the Americas, one of which occurs in Costa Rica.

## Double-striped Thick-knee (*Burhinus bistriatus*)

20 in (50 cm). Bold stripes on head; large yellow eyes and short stout bill; long yellow legs. Light-brown plumage has coarse streaking. Erect posture. Resident; common in Guanacaste and in central Pacific lowlands. Prefers open expanses of sand or dirt with short-vegetation ground cover. Usually crouches or runs away when alarmed, but if flushed from cover flies low with shallow, stiff wingbeats. Mostly nocturnal. During the day, rests in shade—standing or squatting—usually in pairs or small groups. Eats a variety of invertebrates, small reptiles, frogs, and, occasionally, seeds. Call is a long series of rapid clucking notes, loud and far carrying. In some other countries, this species is sometimes domesticated to serve as a watchdog at night.

**Family CHARADRIIDAE (Plovers).** Of the nine species in Costa Rica, several are breeding residents, and several are migrants (and two species have both breeding and migratory populations). Plovers are birds of open fields and shores. Sexes are alike. In migrant species, breeding and nonbreeding plumages tend to be distinctly different. Generally, plovers are very similar to sandpipers.

## Black-bellied Plover (*Pluvialis squatarola*)

12 in (30 cm). In Costa Rica, this large, plump shorebird is one of the most widespread—and easily identified—plovers. Bill is short but stout; in flight, black axillaries are diagnostic. In nonbreeding plumage (Oct. to March), birds are mottled gray above, paler below. In breeding plumage (April to Sept.), face, throat, and breast are black; head and neck white. Both migrants and winter residents are common along the Pacific coast, less common on the Caribbean coast. This species is uncommon in the summer. Found on beaches, mudflats, estuaries, and salt farms; inland it occasionally occurs in open wetlands. Forages alone; congregates in groups when resting during high tide. Call is a sweet, slurred whistle: *peeeowee*.

nonbreeding

breeding

nonbreeding

**Family RECURVIROSTRIDAE (Avocets and Stilts).** There are two species in Costa Rica: the Black-necked Stilt is a common breeding resident; the American Avocet is a rare migrant (not featured).

## Black-necked Stilt *(Himantopus mexicanus)*

15 in (38 cm). An extremely thin bird; long legs and neck—and a needle-like bill—give it a frail aspect. White below, solid black above; bright-red legs. Permanent residents: abundant in lowlands around the Gulf of Nicoya and in Caño Negro region. Migrants and winter residents are less common: found in lowland areas of both coasts. Prefers quiet shallow ponds, marshes, and impoundments, sometimes estuaries; also at river margins. Always in groups, small or large. Wades into water and pecks at the surface to catch small aquatic invertebrates. When alarmed, utters a constant chatter of sharp, nasal *keek* notes. Spanish name is *soldadito* (little soldier).

**Family JACANIDAE (Jacanas).** There are two species in Costa Rica: the Northern Jacana, a breeding resident, is widespread; the Wattled Jacana (not featured) occasionally wanders up from Panama.

## Northern Jacana *(Jacana spinosa)*

9 in (23 cm). Adult has chestnut body; black head and neck; and yellow frontal shield and bill. Young tan above, darker on back of neck and crown; white below; white eye line; frontal shield reduced or absent. In all ages: extremely long toes and nails; open wings show pale yellow. Flies with alternating bursts of quick flaps and short glides. Breeding resident; common countrywide at low and middle elevations; occurs in ponds, marshes, wet pastures, and other areas with still water and floating vegetation (on which the birds walk). Usually in pairs or loose family groups. Eats small aquatic animals and plants. Females mate with several males, each of whom cares for a separate clutch of eggs.

adult

young

**Family SCOLOPACIDAE (Sandpipers).** Twenty-nine species reported in Costa Rica, all migrants or nonbreeding residents. Birds of coastal areas or open wetlands that sometimes gather in large flocks. The majority of species breed in the Arctic (and have brightly colored breeding plumages rarely seen in Costa Rica). Identification is often a challenge. Ages and sexes are similar.

## Spotted Sandpiper (*Actitis macularius*)

breeding

nonbreeding

7.5 in (19 cm). Stocky, with short legs and neck; tends to lean forward, with head held down; habitually teeters. Brownish above, white below except for brownish smudge on edge of breast; pale eye stripe. On breeding birds (April to Aug.), note black spots below. Most widespread shorebird in CR; found countrywide, except in highlands, in almost any aquatic habitat. Migrant and nonbreeding resident; common in most months but rare in summer. Solitary and territorial. Distinctive flight pattern: intersperses rapid, shallow wingbeats (below the plane of the body) with short glides. When foraging, walks briskly, with frequent pauses. Vocal, especially when two birds come too close together.

## Willet (*Tringa semipalmata*)

15 in (38 cm). Erect posture. Large, with somewhat stocky build; legs are grayish and fairly long; bill straight with a dark tip. In flight, shows bold black-and-white pattern on wings. Populations of both migrants and year-round nonbreeding residents; common on Pacific coast, with fewer numbers on Caribbean (though widespread, seldom occurs far from coasts). Prefers beaches, mudflats, estuaries, and sand bars in large rivers. Usually found singly (of the large, generally solitary, shorebirds, this species and the Whimbrel are the two most often seen on beaches); when foraging, sometimes forms loose groups. At high tide, gathers in flocks to rest in salt farms and mangroves. Generally noisy and jumpy; easily disturbed.

nonbreeding

nonbreeding

## Whimbrel (*Numenius phaeopus*)

17 in (43 cm). Large. Long, narrow, decurved bill; striped head. Upperparts grayish brown with light flecks; buff below, with faint stripes on breast and bars on flanks. Populations of both migrants and year-round nonbreeding residents; common on Pacific coast, less numerous on Caribbean. Found on sandy and rocky beaches, mudflats, and estuaries; sometimes occurs in flooded or muddy fields. Usually forages alone, but at high tide forms large flocks that rest in salt farms and mangroves. Walks slowly, picking and poking in search of crustaceans, worms, and other small invertebrates. Often seen with prey grasped in tip of bill.

**Family LARIDAE (Gulls and Terns).** Twenty-two species in Costa Rica; the 5 species that breed in the country all occur on islands in the Pacific Ocean. These coastal and open-ocean birds rarely venture inland. Sexes are alike; young often pass through several plumages before reaching adult form.

## Laughing Gull (*Leucophaeus atricilla*)

16 in (40 cm). When not breeding (Sept. to March), adult has white body and tail, gray back and wings (with black wing tips); head shows dark smudges; bill and legs black. When breeding, head is black, bill and legs are reddish. Young are dirty looking, extensively gray or brown; with dark wings and black band on tail. (Similar Franklin's Gull is smaller; has white on wing tips; and is common only along Pacific coast during migration.) Slow, graceful flight with deep wingbeats. Populations of both migrants and nonbreeding residents are common along both coasts, but more abundant on Pacific coast, especially the Gulf of Nicoya (in Costa Rica, this species is the gull you are most likely to see). Frequents ports, fishing villages, river mouths, beaches, salt farms, and aquaculture farms; on rare occasions, occurs inland, along large rivers. Follows ferries and fishing boats (but only when vessels are close to shore).

first winter

breeding

nonbreeding

breeding

## Royal Tern (*Thalasseus maximus*)

19 in (48 cm). A large tern, with a stout, orange bill and a slightly forked tail; short, black legs. Adult white, with pale-gray back and wings; head crested at rear. When breeding (March to June), note sleek, black cap; in other months, forehead is white. Young similar but with light-brown marks on back, dark bar on wing, and paler bill and legs. Direct, steady flight with deep wingbeats. Populations of both migrants and nonbreeding residents are common year-round, on both coasts. Dispersed groups or lone birds forage over the ocean, close to shore; cruises high up, then plunges down to catch fish at surface of water. Also forages at shrimp-farm ponds. Rests in larger flocks—often with other terns and gulls—at river mouths and on beaches, mudflats, sandbars, and salt farms.

nonbreeding

breeding

**Family COLUMBIDAE (Pigeons and Doves).** Twenty-five species in Costa Rica. Lacking a strong bill, pigeons must ingest pebbles; the grinding action of the small stones facilitates the gizzard's work of digesting hard seeds. Young are fed exclusively on a regurgitated liquid called pigeon's milk, a substance unique to this family. Ages and sexes similar.

## Pale-vented Pigeon (*Patagioenas cayennensis*)

12 in (30 cm). Pale belly, tail, and undertail; rufous shoulders; gray head; entirely dark bill. Breeding resident; common in lowlands of the Caribbean and S. Pacific. Frequents treetops in open and semi-open areas; often near water and especially common along the coast. Alone or in pairs, at times in loose flocks of a dozen or so. Eats berries. In display flight, male climbs with exaggeratedly deep wingbeats, then circles down in a glide, with wings held high. Song a mournful cooing separated by one or two short stutters; aggressive call an abrupt, spitting growl. One of a handful of birds that is resident on Caño Island.

## Red-billed Pigeon (*Patagioenas flavirostris*)

12 in (30 cm). Dusty purplish above; ruddy patch on shoulder; gray wings, rump, and belly; dark tail. Bill white to pink, with an inconspicuous red base. Wingbeats somewhat floppy but flight is powerful and fast. Resident; common in Central Valley, most of Guanacaste, and northern half of Caribbean slope, mostly at middle elevations; less common in lowlands. Prefers open or semi-open spaces with tall trees; forest edges; and agricultural areas. May form flocks of up to 50 birds, though usually seen alone or in pairs. Eats fruits and berries. Song a mournful cooing separated by three short stutters. Spanish name is *paloma morada* (purple pigeon).

## Band-tailed Pigeon (*Patagioenas fasciata*)

14 in (34 cm). Large, gray pigeon; yellowish bill and legs; white crescent on nape. Has a relatively long tail with a wide, pale-gray tip. Resident; common to abundant countrywide in highlands; seasonally descends to foothills, and, rarely, to adjacent lowlands. Prefers canopy of tall forest, where it feeds on fruits and acorns; along quiet roadsides it is occasionally seen drinking from puddles or picking up grit. Occurs in shy flocks of up to 50 birds. Flies high and fast over valleys; also flies low through mountain passes, where it is often hunted. A big flock rushing just above the treetops produces an impressive airplanelike whooshing as it passes overhead. Birds also clap their wings when flushed. Call is a deep, hoarse, hooting *cu-WHOOO*, repeated up to a dozen times.

## Short-billed Pigeon (*Patagioenas nigrirostris*)

10.5 in (26.5 cm). Small pigeon; reddish brown; note short dark bill and reddish eyes and legs. Resident; common on Caribbean and S. Pacific slopes, at low and middle elevations. Occurs in canopy of wet forests; also found in the crown of tall trees in semi-open areas adjacent to forest. Sometimes descends from trees to feed and to pick up grit on roads; eats fruits and berries. Travels alone or in pairs. Far-carrying and leisurely repeated song is a musical, resonant cooing that sounds something like: *who COOKS for YOU*. Call is a soft rolling growl. Spanish name *dos tontos son* (in English, there go two fools) is a mnemonic for its song.

### White-winged Dove (*Zenaida asiatica*)

10.75 in (27 cm). Medium size. Pale brown with white wing bar; long, rounded tail has a white tip. Fast, direct flight with deliberate, pumping wingbeats; takes flight with a whining rustle of wings. Breeding resident: common in dry N.W. Pacific, central Pacific, and Central Valley. Winter migrant: sporadically abundant in Guanacaste, rarely farther south or east. Occurs in open and semi-open agricultural land, pastures, mangroves, and residential areas with tall trees. Often in flocks, sometimes large ones; may nest in low-density colonies. Eats seeds, grains, and fruits; feeds on the ground or up in vegetation. Song a long and lilting series of variably phrased *coos*.

note white on tail and wing

wing

### White-tipped Dove (*Leptotila verreauxi*)

10.25 in (26 cm). Robust terrestrial dove. Back is pale brown; head and underparts are pale gray; note wide white tip on rounded tail—visible when bird flushes. Resident; common on Pacific slope at low and middle elevations; locally common on Caribbean slope at middle elevations (has also begun to colonize deforested lowlands in the Caribbean). Inhabits dry-forest interiors, open second-growth forests, woodlands, coffee fields, and shaded yards and gardens. Usually seen alone, sometimes in pairs. Eats seeds and insects. Walks steadily, veering slightly from one side to the other, with head bobbing. When alarmed, walks away or flies to low, concealed perch (once on perch, it dips head down and simultaneously jerks tail up, repeatedly). Sings when perched (at low and middle heights of trees); song is a low, mournful, drawn-out note, with special emphasis given to the middle of the phrase.

note white tips on tail

### Inca Dove (*Columbina inca*)

8 in (20 cm). Small terrestrial dove; sandy plumage with scaling; long tail has white sides; in flight shows rufous in wings. Resident, common in northern half of the Pacific slope and in the Central Valley. Moved into Costa Rica in the early 1900s, extending the southern limit of its range (appears to be moving southward). Generally occurs in open areas with sparse vegetation: scrubby pastures, agricultural lands, roadsides, and yards and gardens (both rural and urban). Often seen perched on utility wires, usually in pairs or small flocks. Picks grains and seeds from the ground. Flushed birds rattle their wings. From elevated perches, males deliver a slightly raspy two-note song: *who-pu.*

### Common Ground-Dove (*Columbina passerina*)

6.25 in (16 cm). Petite; short wings and tail. Grayish brown; rufous color on wings; finely scaled on head and breast; reddish base of bill. On male, face and breast have a pinkish wash. Resident; common to abundant in dry N.W. Pacific and in the Central Valley. Less common southward on the Pacific coast; also less common in N. Caribbean lowlands. Often found on patches of bare dirt (or sand) in open areas: cultivated fields, scrubby pastures, roadsides, and yards of rural dwellings. Occurs in pairs; less likely to flock than other ground-doves. Walks quickly, with a bobbing motion. Takes flight suddenly on lightly whirring wings. Forages on ground for grass and weed seeds; also eats the occasional small berry or insect. Repetitive call is a mellow (somewhat rising) *huwoo.*

### Ruddy Ground-Dove (*Columbina talpacoti*)

6.5 in (16.5 cm). Small. Male ruddy with gray head; female brownish, with paler head. Resident; common to abundant on Caribbean and S. Pacific slopes, at low and middle elevations. Occurs less commonly in Guanacaste and on outer Nicoya Peninsula. Found in second-growth clearings, agricultural areas, pastures, and yards. Often seen along gravel roads, usually in small flocks. Forages on ground for seeds and small berries. Call is a low, two-noted *per-whoop* (repeated but not incessantly).

male

female

**Family PSITTACIDAE (Macaws, Parrots, and Parakeets).** Seventeen species in Costa Rica, all resident. Powerful hooked bill is used to cut into tough seeds and fruit, and as an aid for climbing. Pairs fly side by side, even within flocks. Males, females, and subadults are generally similar in appearance. Hunted for the pet trade, especially the large colorful macaws.

## Crimson-fronted Parakeet (*Aratinga finschi*)

11 in (28 cm). Largest parakeet in Costa Rica. Long, pointed tail; mostly green plumage, but in flight note red and yellow on underwing; red forehead, pale bill. Widespread and common, at low and middle elevations everywhere except on the Nicoya Peninsula and in N.W. Pacific lowlands. Endemic to Nicaragua, Costa Rica, Panama. This parakeet is the one seen in downtown San José. Favors open country with tall trees, rare in extensively forested zones. Forms large noisy chattering flocks seen flying high or circling over feeding or roosting sites; often roosts in tall palms or clumps of giant bamboo. Eats fruits and flowers; sometimes becomes an agricultural pest. Seldom caged, the volume of its screeching vocalizations is probably unbearable in an enclosed space.

## Orange-fronted Parakeet (*Aratinga canicularis*)

9 in (23 cm). Medium size. Green, with blue and yellow on wings, orange patch above bill, and a pale, prominent eye ring; pointed tail. Common in N.W. Pacific lowlands and foothills; less common farther south and in the western Central Valley. Occurs in forest canopy and in open areas with scattered trees. Flies swiftly, following closely the contour of vegetation. Flocks are small but birds sometimes congregate at feeding and roosting sites. Eats fruits, flowers, and some seeds. Favored nest site is a cavity that it carves into an active arboreal termite nest. Calls screeching, at times almost squeaky. A popular pet.

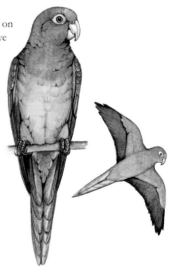

## Olive-throated Parakeet (*Aratinga nana*)

9 in (22.5 cm). Medium size. Green plumage; dull-olive throat, bluish wings; pointed tail. Fairly common throughout Caribbean lowlands, in forests and adjacent open areas with tall trees. Typically in small, fast-flying, compact flocks of 15 to 30 birds. Eats treetop fruits, especially figs and *Inga*. Excavates nest cavities in arboreal termite nests. Vocalizations dry and rattling; less squawky than other parakeets.

## Orange-chinned Parakeet (*Brotogeris jugularis*)

7 in (18 cm). Small; the only smaller Costa Rican parakeet is the rare, nomadic Barred Parakeet of highlands. Big head, small bill, and short, pointy tail; green with brownish patch on wing. The most common of the small parakeets; occurs countrywide at low and middle elevations. Frequents a variety of habitats, but seems to prefer forest edge or open areas with tall trees; except in dry N.W. Pacific, rarely ventures deep within forests. Flies with rapid, stuttering wingbeats, either in pairs or fast-flying compact flocks of 50 or so birds (but even within flocks, segregates loosely into pairs). Eats small fruits and flowers, and some nectar. Calls almost constantly in flight and when feeding; notes are chatty and scratchy, sometimes tinkling. A favorite pet; often seen carried on a stick or riding a child's shoulder.

### Brown-hooded Parrot (*Pyrilia haematotis*)

8.25 in (21 cm). Small and chunky; green plumage, olive-brown head, red ear patch, and a pale eye ring; short, square tail. Red axillaries visible in flight. Common; occurs on Caribbean slope and S. Pacific, in lowlands and foothills. Prefers canopy of forested areas but also found in nearby semi-open areas with tall trees. High-flying flocks are occasionally seen traversing deforested stretches. Fast flying with rapid wingbeats; flaps wings in a wide arc, from high above body to well below (cf. White-crowned Parrot). Flies in tight flocks of 5 to 15 birds (not segregated into pairs). Flocks feed quietly, but explode into noisy flight if startled; consumes a wide variety of fruits and seeds. When perched utters single, inquisitive chirps and also richer, more melodious, notes; calls in flight are sharper, higher pitched, and somewhat metallic.

### White-crowned Parrot (*Pionus senilis*)

9.5 in (24 cm). Medium size; has blue-green body, white crown, red undertail. Common; occurs on Caribbean and S. Pacific slopes, at low and middle elevations (overlaps with similar Blue-headed Parrot near Panama border, on both coasts). Found in canopy of forest, forest edge, and in scattered trees of semi-open areas. Flight typical of genus *Pionus*: slow wingbeats; flaps wings below the body, not above (cf. Brown-hooded Parrot). Sometimes in pairs but more often in flocks of 15 to 50 birds (within flocks, pairs segregate loosely). Often perches in plain view in the tops of tall trees or palms. Eats fruits and seeds (of corn, pejibaye palm, and other crops). Call is a dry, raspy squawk.

### White-fronted Parrot (*Amazona albifrons*)

10 in (25 cm). Smallest of the four *Amazona* species in Costa Rica. Green with fine dark scaling; has white forehead and bluish crown; note red around eye and on small section of bill; in flight shows blue on wing. Male has red on outer wing, female lacks red on wing (and is duller overall). Common; occurs in N.W. Pacific lowlands and foothills; rarely ventures south to Carara or west into Central Valley. Found in mangroves, in canopy of forests, and in tall trees in semi-open areas. Rapid flight with fast wingbeats; generally flies in noisy flocks (less often in pairs) of several-dozen birds, usually not far above vegetation. Eats fruits, flowers, and seeds. Shrill in flight; squawky and more conversational when perched; most common call is a series of *yak* notes.

adult male

young

adult male

### Red-lored Parrot (*Amazona autumnalis*)

13.5 in (34 cm). Fairly large. Green plumage, red forehead, and small red patch on wing. The most common of the three large *Amazona* species in CR; widespread on both Caribbean slope and Pacific slope (S. Nicoya Peninsula and S. Pacific), in lowlands and foothills; Pacific populations show pronounced seasonal movements. (Substantial overlap with similar Mealy Parrot, but latter generally restricted to forested areas.) Flight typical of larger *Amazona* parrots: steady, rather slow flight on horizontal wings that appear to move only at the tips. Alone or in flocks, pairs fly side by side with wings almost touching. Prefers low-density forests or semi-open areas with tall trees; less often in dense forest. Often perches in plain view in treetops. Makes long daily commute between roosting and feeding sites; eats fruits and seeds. Varied calls have a conversational quality; listen for a *ka-link* phrase (unique to Red-lored). Commonly kept as a pet.

### Yellow-naped Parrot (*Amazona auropalliata*)

13.75 in (35 cm). Green plumage; yellow crescent on nape and red patch on wings. Uncommon; occurs in N.W. Pacific lowlands and south to Carara; rare outside protected areas. (Range overlaps with that of much smaller White-fronted Parrot.) Found in canopy of trees in forests and open countryside. In slow but steady flight, pairs fly side by side, wings almost touching (generally in pairs or small family groups; sadly, this parrot is too rare to form large flocks). Eats fruits, seeds, flowers, and buds. Produces a diverse set of vocalizations; generally makes rich, rolling notes that are much less harsh than those of other parrots; at twilight, vocalizations of flying pairs have an eerie quality. A skilled mime of the human voice, and so the most frequently caged *Amazona* in Costa Rica.

**Great Green Macaw** (*Ara ambiguus*)

31 in (79 cm). Very large. Green body; wings bluish above, yellowish below; flash of red in tail (long and pointed); black thick bill. Increasingly rare and local in Caribbean lowlands and foothills; in forested and semi-open areas. Travels great distances to find fruits of tall trees, especially the Wild Almond (*Dipteryx panamensis*), on which it is critically dependent. Usually in pairs or threes; rarely in flocks of a dozen or so. Steady flight with slow, shallow wingbeats. Loud, roaring squawks can be heard from afar; vocal in flight but usually quiet when perched or eating. Endangered, mostly due to habitat loss, but also because of hunting and live capture for the illegal pet trade. With its low rate of reproduction, depleted populations are slow to recover. *Illustration not to scale.*

**Scarlet Macaw** (*Ara macao*)

33 in (84 cm). Very large. Bright red body; blue and yellow on wings; its massive, hooked bill is white above, black below; long pointed tail. The two main populations are confined to Carara and the Osa Peninsula; a few pairs in Guanacaste and N. Caribbean lowlands. Found in forest canopy and in trees in semi-open areas. Eats large fruits and seeds; sometimes seen at beaches eating fruits of the Tropical Almond (*Terminalia catappa*). Mostly occurs in small groups, forming larger flocks when roosting. Often makes long daily commutes between roosting and feeding sites. Steady flight with slow, shallow wingbeats. Quiet when eating on perch, but vocal in flight; loud, roaring squawks are far carrying. Conservation efforts—including nest protection, creation of artificial nests, and captive breeding and release programs—have helped offset damage done by illegal pet trade. *Illustration not to scale.*

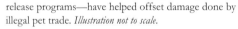

## Order Cuculiformes (Cuckoos and relatives)

**Family CUCULIDAE (Cuckoos).** A diverse family though all have long, often showy, tails; stout bills; and strong feet with two toes facing forward and two backward. Twelve species recorded in Costa Rica; of these, 9 are resident, 3 are migrants. (The Cocos Cuckoo is endemic to distant Cocos Island.)

### Squirrel Cuckoo (*Piaya cayana*)

18 in (46 cm). Slender with a long striking tail. Chestnut upperparts shade to gray on lower breast; underside of tail has bold black-and-white pattern; bill and eye rings yellowish green, eyes red. Resident; common countrywide from lowlands to highlands. A solitary, furtive inhabitant of trees, shrubs, and tangles in both forests and open country. Runs and jumps from limb to limb—never opening wings—which lends a squirrel-like impression to its movements. Eats insects, small lizards, and caterpillars (spiny or hairy caterpillars are beaten against a branch before they are swallowed). Calls are often best clue to its presence; makes a *chick*, *burr* sound suggestive of a distant wolf whistle and a sneering *ricky-here*, both at long intervals; also whistles a long, slow series of *wick* notes.

### Striped Cuckoo (*Tapera naevia*)

12 in (30 cm). Has a bushy crest that it habitually raises and lowers. Brown streaked with black above, buff white below; tail long; bill short and decurved; note white line above the eye. Resident; common countrywide in lowlands and foothills, though rare in dry N.W. Pacific. Inhabits low second growth, brushy clearings, and overgrown fields, where it hunts for insects on the ground and in vegetation. Hard to detect unless singing, when it moves to high open perches (fence posts and power lines, e.g.). Most common call is a clear, far-carrying, two-note whistle that it repeats incessantly (first note low, the second a halftone higher). Like Old World cuckoos, this species (and the similar Pheasant Cuckoo) is a brood parasite: it lays its eggs in the nest of another bird—typically domed nests of spinetails, wrens, or sparrows—thus delegating the task of rearing its young.

### Groove-billed Ani (*Crotophaga sulcirostris*)

12 in (30 cm). All black; long floppy tail; its velociraptor-like bill is lightly grooved. Resident; common countrywide (except in S. Pacific), in lowlands and, locally, at middle elevations. (The very similar Smooth-billed Ani occurs in the S. Pacific; the Smooth-billed Ani has a larger bill that extends above the line of its forehead.) Prefers scrubby second growth, pastures, plantations, and open woodlands. Flight weak: several quick flaps alternate with long unsteady glides. From the ground it often follows livestock, hopping and dashing after the insect prey the animals stir up. A highly social bird; when perched, individuals pack together on a limb, sometimes climbing over each other and even preening neighbors. A communal breeder: several females lay eggs in a single nest that is cared for by all. Excited call is a repeated *TEE-ho*.

**Family TYTONIDAE (Barn Owls).** This mostly Old World family contains 11 species. Its sole representative in Costa Rica is the Barn Owl (*Tyto alba*). Birds of this family are distinguished from owls in the family Strigidae by long feathered legs, a heart-shaped facial disk, comblike serrations on the central claw, and the absence of ear tufts.

## Barn Owl (*Tyto alba*)

16 in (40 cm). Almost entirely white below, tawny to gray above, with fine black flecks over most of body; small dark eyes, white-to-buff face. Females are generally darker overall and have more mottling below. Resident; uncommon but occurs countrywide in agricultural areas and towns. Strictly nocturnal. Nests and roosts in tree cavities, rock formations, old buildings, church steeples, and under bridges. Hunts on the wing, detecting prey with vision or acute hearing. In flight utters both a drawn-out, raspy shriek and a weird clicking sound (but never hoots).

**Family STRIGIDAE (Owls).** Although 16 species have been recorded in Costa Rica, it is unlikely that either the Burrowing Owl or the Short-eared Owl (both rare migrants) occurs here today. Sexes share similar plumage but in most species females are larger than males. Downy-soft plumage permits noiseless flight. Owls swallow prey whole, then regurgitate pellets that contain indigestible parts.

## Spectacled Owl (*Pulsatrix perspicillata*)

19 in (48 cm). Large and distinctive. Bold marks on face, large yellow eyes, and no ear tufts. Belly is buff; wings (with faint barring) and back are brown. Young have fuzzy appearance; entirely white except for black face and brown wings and tail. Resident; fairly common countrywide at low and middle elevations. Lives in forests but also occurs in adjacent open countryside. Mostly nocturnal; at night—from a medium-high perch—hunts mammals, birds, and large insects. After hunting, returns to the same day-roost site, usually situated rather low in dense vegetation; if flushed during the day, the owl is quickly mobbed by smaller birds attempting to drive it away. Vocalization is a series of deep pulsating hoots that start slow but build rapidly in tempo. *Illustration not to scale.*

### Pacific Screech-Owl (*Megascops cooperi*)

9 in (23 cm). Medium size. Cryptically patterned: gray plumage with white speckles and black streaks. Yellow eyes; ear tufts, when held flat, show as mere bumps; feathered legs. Of the four species of screech-owl in Costa Rica, this is the only species whose toes are covered with bristles; the other three screech-owls have scaly toes. It is also the only one that occurs in the dry N.W. Pacific, where it is quite common. Inhabits forests, woodlands, scrubby savanna, mangroves, and open agricultural areas that contain trees. Roosts in tree cavities or out in the open, relying on cryptic plumage and frozen posture for concealment. At night usually hunts large insects from a low perch. Song consists of a slightly accelerating series of 10 to 15 hoots with a barking quality.

### Ferruginous Pygmy-Owl (*Glaucidium brasilianum*)

6 in (15 cm). Tiny. Rufous above, streaked below; false eyespots on back of head; has a longish tail that it frequently cocks and then twitches from side to side. Resident; common in N.W. Pacific, extending south to Parita (and east to beyond Cartago), at low and middle elevations. No overlap with Costa Rican Pygmy-Owl (in highlands) and Central American Pygmy-Owl (on Caribbean slope). Occurs in dry forests, open woodlands, and agricultural and residential areas with tall trees. Active mostly at dawn and dusk. Eats insects, lizards, and birds that sometimes equal it in size. Nests in abandoned woodpecker holes. May call at any time of day or night, whistling a long series of evenly spaced toots.

### Black-and-white Owl (*Ciccaba nigrolineata*)

15 in (38 cm). A fairly large owl. Lacks ear tufts; dark eyes, black facial disks; blackish gray above, *appears* pale gray below (fine black-and-white barring on underparts is hard to discern except at close range). Resident; uncommon but occurs countrywide in forests of low and middle elevations (relatively more common in humid regions). Prefers upper levels of forests but also frequents forest edges and tall trees in nearby open areas. Strictly nocturnal; often stakes out artificial lights to capture the large insects and bats that they attract. Calls include single hoots, wailing screams, and a short series of rising hoots that ends with a single emphatic hoot. Young birds begging for food emit an intermittent, hissing screech that can continue all night.

**Family CAPRIMULGIDAE (Nightjars).** Ten species recorded in Costa Rica, two occur only as migrants. Nocturnal birds with cryptic plumage, and thus difficult to identify except by call. Most species are poorly known. All have small weak feet and tiny bills; they catch flying insects with a huge, gaping mouth.

## Common Pauraque (*Nyctidromus albicollis*)

11 in (28 cm). Medium size. In repose, wing tips extend halfway down tail (wing tips extend farther in many other nightjars). On male, white outer tail feathers are diagnostic; female has white-tipped tail. Red eyeshine. (Aside from the three nighthawk species, this is the only nightjar encountered with any regularity in CR.) Resident; common (to abundant) countrywide, in lowlands and, locally, middle elevations. Occurs in a variety of semi-open areas, both brushy and woody. At night, most often seen perched in the middle of a dirt road; during the day, usually seen when flushed from its thicket roost on the ground. Sallies from ground to catch flying beetles and moths. Lays eggs in ground nest; parents attempt to distract predators from nest by performing a broken-wing display. Very vocal; typical call is a whistled *cur-WEE-o*, tirelessly repeated (often heard on moonlit nights).

male

male

**Family NYCTIBIIDAE (Potoos).** This family occurs only in the New World tropics. Three species in Costa Rica. The Common Potoo, p. 92, and the Northern Potoo are virtually identical. Nocturnal birds that are often heard but seldom seen. During the day, cryptic plumage and a stretched-out pose make roosting birds difficult to distinguish from broken-branch stubs. Potoos look more owl-like at night, when they open their huge eyes.

## Great Potoo (*Nyctibius grandis*)

20 in (51 cm). Large bird. Long full tail and long wings; gray-buff plumage is cryptically patterned. Sexes are alike. Resident; uncommon in lowlands of both Caribbean slope and the Osa Peninsula. Occurs in canopy of mature forests and in tall trees of nearby semi-open areas. By day, roosts high in canopy, motionless, often on an open branch. At night, hunts from high exposed perch; sallies out to catch with its gaping mouth large flying insects and even small bats, then returns to same perch. Nestles its single egg in crook of branch or on top of dead snag. When perched gives a deep throaty roaring *GWAAAAA*; in flight, emits a higher pitched and more emphatic *GWOK* (calls are far carrying and unearthly). *Illustration not to scale.*

# Living on the Edge

Mother with chick (Common Potoo, *Nyctibius griseus*).

The Common Potoo spends the day perched upright on a stub or branch, relying on its mottled feathers to blend in perfectly with its woody perch. The mother lays a single egg in a shallow depression on a branch or stub—without a nest. Both parents incubate the egg, one during the day, the other throughout the night. When threatened, an incubating parent opens its mouth wide to display the brilliant violet-pink color of the mouth interior, startling potential predators. After a month or so, the egg hatches. The whitish, downy chick remains motionless on its perch and is cared for by both parents until it becomes a competent flyer, about 50 days after birth. Dual parenting is rare in most animals but common in birds. (Not described within the species accounts.)

**Family APODIDAE (Swifts).** Eleven species in Costa Rica. Swifts are accomplished fliers; except when nesting, they spend all waking hours on the wing, catching insects in flight, and even mating in the air. They can only take flight when in free-fall; cling to vertical surfaces when roosting. The nest of some species is made from gluey saliva. Identification to species is often challenging.

## White-collared Swift (*Streptoprocne zonaris*)

8.75 in (22 cm). Largest swift in CR. Entirely black except for white collar. Notched tail. Resident; common countrywide at low and middle elevations (though less common in dry N.W. Pacific). Direct flight is fast and powerful, falconlike (compare with Bat Falcon, p. 72); when foraging, flight incorporates more gliding and veering; often soars with wings and tail spread. Usually in flocks of 25 to 50 birds, but sometimes forms tightly wheeling flocks of several hundred. Roosts and nests in mountain sites but may visit either coast during the day. Places nest in rocky crevices that lie behind (or near) waterfalls. Sometimes gathers at the leading edge of storms to collect insects swept into the air by the turbulence. Quite vocal, especially when a pair of birds chase each other or entire flocks break into a noisy screeching chatter.

**Family TROCHILIDAE (Hummingbirds).** This family is confined to the New World. There are 52 species in Costa Rica. Hummingbirds feed on flower nectar: bills of many different shapes and sizes are each adapted to fit a certain form of flower; these birds play an important role as pollinators. Nectar diet is supplemented with insects and spiders that are either caught in the air or plucked from a web. Females attend to all nesting duties and raise young without help from males.

A note on hermit hummingbirds (subfamily PHAETHORNITHINAE): Six species from this distinct subfamily occur in Costa Rica. All reside in forest understory. Hermits are generally characterized by elongated tails and long, very curved bills. They are the principal pollinators of many species of *Heliconia*. The three species described here all form breeding leks.

### Green Hermit (*Phaethornis guy*)

6 in (15 cm). Long, very curved bill is tinted reddish below. Male dark green above; short white tail-spike; in good light, note blue-green rump. Female green above and gray below; pale facial stripes; long white central tail feathers. Common countrywide at middle and upper elevations (the only large hermit that occurs in CR mountains). Inhabits understory of wet forests and nearby in woodlands, plantations, and gardens. Follows an established foraging route, making brief visits at each flower before moving on to the next. At breeding leks males perch knee-high, and, while flicking tails, utter a squeaky chirp, repeated over and over. In addition to calls, males produce loud wing-snaps.

### Long-billed Hermit (*Phaethornis longirostris*)

6 in (15 cm). Sexes and ages similar. Bronzy green above, shading to cinnamon on rump; pale buff below. Curved, very long bill; long, white-tipped central tail feathers; pale facial stripes. Common to abundant in lowlands of Caribbean and S. Pacific; rare in N. Pacific, in mangroves and evergreen forests. Inhabits forest understory and adjacent woodlands, plantations, and gardens. Inquisitive; often zips to within a few feet to inspect birders—maneuvering to stay behind viewer's head, despite one's best efforts to turn and look at it. Follows regular foraging routes that can be a half-mile long. From low perches males tirelessly repeat a single squeaky note, all the while pumping their tail. The nest of the Long-billed (as with most hermits) is attached below the leaf tip of a palm or *Heliconia*. (Formerly named Long-tailed Hermit.)

### Stripe-throated Hermit (*Phaethornis strigularis*)

3.5 in (9 cm). Tiny. Curved bill is of medium length; plumage mostly cinnamon colored but note pale facial stripes and dark wings; tail dark with elongated central tail feathers that are buff-tipped. Common; countrywide at low and middle elevations (in dry N.W. Pacific, restricted to patches of evergreen forest). Occurs in forest understory and adjacent semi-open areas. Visits a variety of small flowers and also pierces base of larger flowers to get at nectar. Male wags tail and sings squeaky song from a very low perch in dense cover (calling birds are very hard to locate); song is more varied than in other hermits. In flight, wings produce a distinctive, buzzy sound that allows identification even if bird is not seen. (Formerly named Little Hermit.)

### Violet Sabrewing (*Campylopterus hemileucurus*)

6 in (15 cm). Largest hummingbird in CR. Bill curved, medium length; full tail with prominent white corners. Male has bright-purple plumage that shades to greenish on back. Female is slightly smaller; greenish above and gray below; purplish throat. Common; occurs throughout CR at high elevations; seasonally, some birds venture down to middle elevations. Found in forest understory, forest edges, and clearings. Visits *Heliconia* flowers, banana plants, and some large ornamental flowers; regular visitor to feeders. Sometimes defends a patch of flowers. Up to a dozen males form a dispersed lek, singing while perched at low to medium heights. Song is a variety of loud chirps and twitters delivered in a slow-cadenced series.

male

female

## White-necked Jacobin (*Florisuga mellivora*)

4.75 in (12 cm). Bill is short, thick, and almost imperceptibly curved. Male has deep-blue head, white crescent on nape, green back, and white underparts and tail. Female usually green above, a scaly green and white below; tail dark with pale tips at the corners (some females resemble males). Fairly common; occurs on Caribbean and S. Pacific at low and middle elevations. Male often perches conspicuously on a high, bare twig; both sexes cock tails when hovering at flowers (male also occasionally fans his tail). On the wing, snatches flying insects. Visits many types of flowering trees, vines, and shrubs in forest, forest edge, and open country with tall trees. Males seen most often in canopy or outside of forest; females seen mostly in forest interior, where their cuplike nest (made from plant fuzz) sits atop a leaf, typically of a short palm.

female

male

## Green Violet-ear (*Colibri thalassinus*)

4 in (10.5 cm). Medium size. Green plumage, violet ear patch, dark subterminal band on bluish, square-tipped tail. Bill slightly decurved. Male shows glittering green on throat, otherwise sexes are similar. Common; occurs at high elevations of Tilarán Mountain Range and south to Panama. Prefers forest edge and clearings. Visits many kinds of flowers, including species on the ground and those high in the canopy; also visits feeders. Males sing from high—often exposed—perch, but can be difficult to locate; song is a loud, alternating *chip, chop* (sometimes *chip-chip, chop*) with a constant rhythm, like the ticking of a clock.

## Blue-throated Goldentail (*Hylocharis eliciae*)

3.5 in (9 cm). Green above; glittering greenish-gold tail; buff belly. Male with blue to violet throat. Bill short, straight, and broad at base; on male bright red with a black tip, on female mostly black. Widespread on both slopes; more common on Pacific, especially in humid lowlands and valleys, but scarcer and local in dry N.W. Pacific and throughout most of the Caribbean lowlands. Prefers forest, forest edges, and shady gardens. Forages mainly at middle level of canopy, taking nectar from flowers of small trees and shrubs. In dry season, males assemble in loose groups in forest interior, and from open perches tirelessly repeat a chipping display-call to attract females.

male

female

## Volcano Hummingbird (*Selasphorus flammula*)

3 in (7.5 cm). Tiny. Short straight bill; central tail feathers green to blackish. Male green above, pale below (with greenish-gray flanks); color on male's throat varies geographically, from rose red in northern populations to purplish green in southern ones. Female is similar, but note speckled throat and tail with black band and pale outer tips. (cf. Scintillant Hummingbird.) Very common; occurs on Central and Talamanca mountain ranges at high elevations. Endemic to Costa Rica and western Panama. Ranges from forests to above tree line (in páramo); prefers disturbed areas (gaps and brushy clearings). Takes nectar from a variety of small flowers; at larger flowers, obtains nectar by using holes previously made by flower-piercing birds and bees. The display flight of the male is a repeated, high, looping "U" performed in a clearing, above favorite perch. Wings produce a twanging sound during display—and male vocalizes a high-pitched twittering.

male
Irazú & Turrialba
volcanoes

male
Poás & Barva
volcanoes

female

male
Talamanca
Mountains

## Scintillant Hummingbird (*Selasphorus scintilla*)

2.5 in (6.5 cm). Smallest hummingbird in Costa Rica (endemic to Costa Rica and Panama). Similar to Volcano Hummingbird but has pronounced white collar, more rufous on belly, and a markedly rufous tail that lacks green (or black) central tail feathers. Male's throat orange-red. Common; occurs from Tilarán Mountain Range south to Panama, at middle and high elevations (generally at lower elevations than Volcano Hummingbird). Visits low-growing flowers at forest edges, brushy areas, pastures, plantations, and gardens. Adept at stealing nectar from flowers defended by larger hummingbirds. Male has a looping display flight similar to that of male Volcano Hummingbird. Most notable sound during normal flight is a metallic buzzing produced by wings.

male

female

### Rufous-tailed Hummingbird (*Amazilia tzacatl*)

4 in (10 cm). Medium size. Body green (brightest on throat and breast), belly grayish, tail rufous. Bill straight, medium length; male's bill is red with a short black tip; on female lower mandible is red, upper mandible red at base with a long black tip. In CR, this is the most common hummingbird in lowlands, except in N.W. Pacific, where it is often replaced by Cinnamon Hummingbird. Prefers open areas of both rural and urban settings, but will also enter forest. Visits many types of flowers; highly territorial, it dominates other hummingbirds at flowers and feeders. Nest is a compact cup of fine grayish fibers bound by spider webs and decorated outside with flecks of lichen. Male song is a simple, repeated pattern of 3 to 5 squeaky notes, usually sung during first hour of the morning only; both sexes call frequently when foraging.

male

### Cinnamon Hummingbird (*Amazilia rutila*)

3.75 in (9.5 cm). Medium size. Bronzy green above, cinnamon below, tail rufous. Bill straight, medium length; lower mandible is red, color of upper mandible varies depending on age and sex. Common; occurs in N.W. Pacific south to Orotina (and into western part of Central Valley), in lowlands and foothills. Occasionally wanders far outside its range. Found in dry forest, woodland and open scrub, agricultural areas, and gardens. Replaces Rufous-tailed Hummingbird in N.W. Pacific. Like the Rufous-tailed, it is aggressive and territorial at flowers; it visits a wide variety of flowers at all levels of forest. Calls are dry rattling notes that are sometimes confusingly similar to call of Banded Wren (ranges do overlap).

## Coppery-headed Emerald (*Elvira cupreiceps*)

3 in (7.5 cm). Small. Bill short and curved. Male appears bronzy green above, with a coppery tone on crown, rump, and upper side of tail; glittering green below except for thighs (white) and undertail; thin black tips on outer tail feathers. On female, green upperparts lack coppery iridescence; whitish below; relatively broad black tips on white outer tail feathers. Female Black-bellied Hummingbird is similar but note straight bill and rufous patch on wing (little range overlap). Restricted to a narrow strip of middle-elevation wet forest that runs from Guanacaste to the slopes of Turrialba Volcano. Found in forests (at all levels), forest edges, and nearby clearings. Moves hastily between small flowers; also regularly visits hummingbird feeders. This is one of only three species endemic exclusively to mainland Costa Rica.

## Purple-crowned Fairy (*Heliothryx barroti*)

4.5 in (11.5 cm). Bright green above, pure white below; black mask; short, straight, needle-tipped bill; long graduated tail (female's tail is longer). Male has violet crown. Uncommon; occurs on Caribbean slope and in S. Pacific at low and middle elevations. Occupies middle to upper levels of forest, open woodland, gardens, and plantations. Makes short darting flights from one spot to another, hovering briefly, with rather slow wingbeats; shows a flash of white as it opens and closes its tail. Often flies in close to check out observer. Hawks flying insects; visits variety of flowers, some of which it pierces at the base to obtain nectar.

## Purple-throated Mountain-gem (*Lampornis calolaema*)

4 in (10.5 cm). Medium size. Green above; dark cheek; long, white eye stripe; straight bill. Male has shiny turquoise crown; purple throat (rest of underparts grayish green); dark-blue tail. Female lacks turquoise crown; entirely rufous below; dull-green tail (with pale-white tips at the corners). Female virtually identical to female White-throated Mountain-gem. Common; endemic to foothills and highlands of S.W. Nicaragua, Costa Rica, and western Panama. In CR, occurs mainly on the Caribbean slope. Favors canopy of cloud forest but may descend to understory at forest edges and clearings. Likes flowers of epiphytes and shrubs. Visits feeders. Aggressive and territorial. (Some authors consider this species and the White-throated Mountain-gem to be color morphs of a single species known as Variable Mountain-gem.)

male

female

## White-throated Mountain-gem
### (*Lampornis castaneoventris*)

4 in (10.5 cm). Medium size. Green above; dark cheeks; long, white eye stripe; straight bill. Male has shiny turquoise crown; white throat (rest of underparts are grayish green); gray tail. Female lacks turquoise crown; entirely rufous below; dull-green tail (with pale-white tips at the corners). Female virtually identical to female Purple-throated Mountain-gem. Common; occurs from Cerro de la Muerte south to Panama, on high- and mid-elevation mountain slopes (endemic to Costa Rica and Panama). Prefers cloud forests (especially when composed of oaks), where it occurs at all levels; also ventures to forest edges and nearby clearings. Feeds on flowers of epiphytes and shrubs, also visits feeders. Sometimes territorial.

male

## Fiery-throated Hummingbird (*Panterpe insignis*)

4.25 in (11 cm). Medium size. Generally appears dark green; white spot behind eye; straight bill; broad, dark-blue tail is slightly notched. Under certain light, iridescent colors are visible on crown (blue) and throat (yellow, orange, and violet). Sexes alike. Common to abundant; occurs at high elevations in all major mountain ranges, from Miravalles Volcano in N.W. Costa Rica to Panama border (endemic to this region). Usually in forest canopy but also lower in clearings and along forest edges. Relatively slow wingbeats for a hummingbird. Visits mostly flowering epiphytes and shrubs; also visits feeders. Pugnacious; defends flowers and will chase away other hummingbirds, Slaty Flowerpiercer (p. 147), and bumblebees. Maintains a constant, harsh, metallic twittering.

throat colors displayed

## Green-crowned Brilliant (*Heliodoxa jacula*)

5 in (13 cm). Large. Tail forked and fairly long; note white spot behind eye; bill is straight and, relative to body size, short. Male dark green; under certain light, note glittering green on crown, throat, and upper breast, and definitive violet-blue spot on throat. Female green above; speckled white and green below; pale tips on tail. Young similar to adult but with cinnamon on face and throat. Common; occurs on all major mountain ranges at high and middle elevations (periodically ranging lower); in northern part of country, mainly found on Caribbean slope, but, moving south, found on both slopes. Prefers canopy and middle levels of wet forest, especially where *Marcgravia* flowers grow. Has unusually strong feet for a hummer and often perches to feed; a frequent visitor at feeders. In flight utters a repeated singsong *chee-chee-chu*.

## Magnificent Hummingbird (*Eugenes fulgens*)

5 in (13 cm). Large and slender; long straight bill; notched tail. Male green above, dark below; purple crown, blue-green throat; white spot behind eye; tail entirely dark. Female smaller than male; green above, gray below; white stripe behind eye; white tips on outer tail. Common; occurs in Central and Talamanca mountain ranges at high elevations (somewhat lower seasonally). Prefers oak forests, where it visits low-growing flowers at forest edges and in clearings. Wingbeats are very slow, almost distinguishable! Call consists of rough, buzzy notes that it often repeats when foraging. (Populations in Costa Rica and Panama are so isolated from those to the north that some authors treat them as a separate species.)

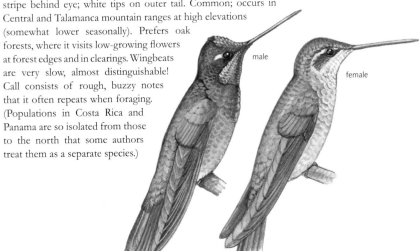

**Family TROGONIDAE (Trogons).** Ten species in Costa Rica, all resident. Trogons are forest dwellers often seen sitting upright on a high branch. Characterized by colorful plumage, big heads, short broad bills, and square tails that are long and broad. Small feet with two toes facing forward and two backward. Males are colorful and often iridescent, females are typically a drab version of male.

## Black-headed Trogon (*Trogon melanocephalus*)

10.5 in (27 cm). Yellow belly; underside of tail has extensive white tips (not barred as in many trogons); note pale-blue eye ring and bill. Male has black head and breast; green-blue back; and tail that shows mostly white below. Female duller; tail banded black and white. Common; occurs in N.W. Pacific south to Carara (and east to Caño Negro), in lowlands. Found in gallery forests (both dry and evergreen) and in semi-open woodland and scrub. Alone or in pairs (rarely in small groups) sits quietly at middle heights then flutters out to snatch fruit or large insects. Often seen traversing open spaces in undulating flight. Nest is a cavity excavated in arboreal termite nest. Song is an accelerated series of whistled clucks that fall in pitch.

## Violaceous Trogon (*Trogon violaceus*)

9 in (23 cm). Small. Yellow belly; underside of tail barred; grayish bill. Male blue above; yellow eye ring. Female dark gray above; grayish eye ring. Found countrywide at low and middle elevations. Common in wetter zones, scarce and local in dry N.W. Pacific, where it is confined to evergreen gallery-forest. Prefers low-density forests, tall second growth, and semi-open areas with tall trees. Alone (or in pairs) perches at middle heights or above. Sallies to pluck fruits and insects from outer branches and foliage. High in a tree excavates its nest in the active nest of wasps, ants, or termites. Song is a long series of notes whistled in same pitch, at unvarying tempo. Birds often sing from high in the canopy and can be maddeningly difficult to locate.

male

female

## Black-throated Trogon (*Trogon rufus*)

9 in (23 cm). Small. Yellow belly; bluish eye ring; greenish-yellow bill; barred tail. Male is green above, has black face and throat. Female is rufous above. Common; occurs on Caribbean and S. Pacific at low and middle elevations. Found in the interior of moist forests, usually in understory (or just above). Alone or in pairs. Nest is placed at about eye level in natural (or excavated) cavities in rotten trunks and snags. In short sallies, plucks fruits, snatches insects off vegetation. Birds often go unnoticed until they flush at close range and emit a startled clucking; once flushed they fly a short distance to perch, from whence they make a rolling *purr*, nervously cocking tail then lowering it slowly. Song consists of three short whistles.

male

female

## Collared Trogon (*Trogon collaris*)

10 in (25 cm). Small. Yellow bill; thin white band above red belly; barred tail. Male is green above, dark eye ring. Female is brown above (grading to rufous on tail), whitish eye ring. (Nearly identical to Orange-bellied Trogon except for color of belly; some overlap.) Common; occurs in Central and Talamanca mountain ranges at middle and high elevations. Inhabits middle levels of wet forest, forest edges, and nearby pastures with tall trees. Alone or in pairs, sometimes in small groups. Nest is a low open cavity carved out of a rotten tree stump or fence post. Adult's body or tail often protrudes from nest cavity. Eats a variety of insects, some fruits. Vocal when in groups; typical song has three whistled notes. Also gives a soft churring call as it jerks tail up and then slowly lowers it.

male

female

## Slaty-tailed Trogon (*Trogon massena*)

12 in (30 cm). Large. Red belly; dark tail (no bars); orangish bill, dark eyes, reddish eye ring. Male green above. Female grayish above. Young have faint barring on tail. Common; occurs on Caribbean and S. Pacific at low and middle elevations. (Overlaps with Lattice-tailed Trogon on Caribbean slope at middle elevations. Lattice-tailed has yellow bill, white eyes, and a finely barred undertail.) Occupies canopy and middle levels of forest, tall second growth, and semi-open areas with tall trees. Usually alone or in loose pairs. Swoops off perch to hover above foliage, then eats fruits, insects, and, occasionally, small lizards or frogs. May join flocks of other large canopy birds or follow monkeys (to eat the prey they stir up). Nest is a chamber excavated inside large arboreal termite nest. Song is a long series of same-pitched husky clucks or barks.

male

female

## Resplendent Quetzal (*Pharomacrus mocinno*)

male

14 in (36 cm); male's tail plumes add 25 in (64 cm) to length. Largest and most spectacular trogon. Male iridescent green to blue-green above; crimson-red belly; thin Mohawk-like crest; four long, green tail plumes; white undertail. Female dull green above; gray breast, red on lower belly; lacks crest and tail plumes; tail with bold black and white barring underneath. Fairly common; occurs from Tilarán Mountain Range south to Panama, in high-elevation wet forests (some movement to lower elevations after breeding). Frequents forest canopy and tall trees in clearings and pastures. Prefers large fruits, especially wild avocados, which it swallows whole (the pits later regurgitated); also eats insects and small vertebrates. Nest is a cavity carved into a large-diameter rotten tree. Startled birds flush with a loud cackling. Male's song is a series of deep, slurred *k'yo, k'yu* whistles that rise and fall in pitch.

female

**Family MOMOTIDAE (Motmots).** Occur only in the New World tropics; six species in Costa Rica. Motmots have a slightly curved bill (with a serrated edge) and small feet. Most have long racquet-tipped tails that they habitually sway, pendulum-like, from side to side. Sexes alike. Tend to perch quietly for long periods at about eye level (or higher).

## Blue-crowned Motmot (*Momotus momota*)

15.5 in (39 cm). Ochraceous olive below; green back, wings, and tail; turquoise blue separates black crown and mask; red eye. Generally common on Pacific slope at low and middle elevations, though rare in N.W. Pacific lowlands. Inhabits forest and shaded plantations and gardens. Solitary or in pairs. From a low perch, drops to the ground to snatch with its bill small vertebrates and insects. Also eats fruit and visits feeders. Nest is a long burrow dug into earthen embankments. Burrow becomes noticeably foul-smelling by the time nestlings are ready to fledge. Of its many vocalizations, the one most often heard is a soft hooting, given either in a series or as a doubled *woot woot*, from which derives the name *motmot* (of indigenous origin).

## Rufous Motmot (*Baryphthengus martii*)

18 in (46 cm). Largest motmot. Rufous on much of body; black mask; green back, wings, and tail (tail long, with small racquets). (Range overlaps with similar Broad-billed Motmot, but latter is smaller and has a blue chin.) Uncommon; occurs on Caribbean slope, in lowlands and foothills. Inhabits understory and middle level of forests; seldom found in deforested landscapes. Usually alone or in pairs. Perches quietly, but sometimes flips around suddenly to face in the opposite direction. Sallies to the ground to catch small animal prey but also plucks prey (and fruit) from vegetation; attends army-ant swarms to eat the prey they stir up. Nest is a chamber at the end of a 12-ft-long burrow dug into a dirt bank. Call consists of deep, resonant hoots (usually given in short bursts). At dawn and dusk chorus (often performed from canopy), birds sing back and forth, alternating *bu,bu,bu* with *ba,ba,ba*.

## Turquoise-browed Motmot (*Eumomota superciliosa*)

13.5 in (34 cm). Small. Pale-turquoise eye brow, black mask and throat stripe, russet back, and blue-green wings and tail. Note broad tail racquets at the end of long bare shafts. Common; occurs in N.W. Pacific and south to Carara, in lowlands and foothills (less common in Caribbean lowlands north of the Guanacaste Mountain Range). Found in a variety of habitats, including forest and scrub. Often perches on electrical wires and fences; remains motionless (except for pendulum motion of tail), then darts out to snatch insect prey from ground or vegetation (also snatches prey in flight). Like other motmots, often smacks prey against perch in an effort to remove indigestible parts. Constructs its burrow nest in embankments. Repertoire of calls includes a *cawak* and a more resonant, rapidly repeated, *parabo*. (The national bird of Nicaragua.)

**Family ALCEDINIDAE (Kingfishers).** All six of the New World members of this family occur in Costa Rica. Five are breeding residents; Belted Kingfisher is a northern migrant. Characterized by large heads (often crested), short necks, large pointed bills, and short legs and tails. Feed primarily on fish, plunge-diving to catch prey with bill.

## Ringed Kingfisher (*Megaceryle torquatus*)

16 in (41 cm). Largest kingfisher in the New World. Bushy crest, oversized bill, white collar. Blue above; male reddish below except for white undertail, female reddish below except for blue and white breast bands. Common; occurs countrywide at low elevations and, occasionally, middle elevations. Found at lakes, rivers, and estuaries (prefers deep, large bodies of water). Alone or in pairs. Tends to perch higher than other CR kingfishers. Often hovers when fishing. Nest is a burrow (with oval-shaped entrance) dug into a high embankment. Produces loud vocalizations. In flight, when high up, repeats single raspy *kleck* note; when agitated, a long, harsh rattle.

male

female

## Amazon Kingfisher (*Chloroceryle amazona*)

11.5 in (29 cm). This is the largest green-backed kingfisher. Dark metallic green above, white collar, almost no white on wings or tail; slight tuft on head; large heavy bill. On male, breast band is rufous; on female, breast band is formed by green spots on a white background. Common; occurs countrywide at low elevations. Fond of slow-moving rivers, streams, and canals (less often at lake shores, estuaries, and in mangroves). Alone or in pairs. Perches fairly low (often in the open) when fishing; sometimes hovers before snatching prey. Carves burrow nest into stream banks. Single birds regularly emit a dry *chrit*, sometimes repeated to create a chattering sound; also produce a series of agitated whistles.

male

female

## Green Kingfisher (*Chloroceryle americana*)

7 in (18 cm). Small. Dark metallic green above; white collar; wings and tail have white spots and bars; white on outer tail (conspicuous in flight). Male has ruddy breast band, female has two indistinct greenish bands. Common; occurs countrywide at low and middle elevations. Found mostly along small wooded streams, but also at edges of rivers, lakes, marshes, and mangroves (even shaded puddles and ditches if these harbor prey). Alone or in pairs. Flight fast and direct, just above surface of water. Perches low, on branches or rocks; when alarmed, nervously cocks tail and lifts head. Eats small fish and some aquatic insects. The entrance to its burrow nest is often screened by vegetation. Call is a single sneezy note; also makes a repeated ticking sound, like pebbles being knocked together.

## American Pygmy Kingfisher (*Chloroceryle aenea*)

5 in (13 cm). Tiny, inconspicuous. Lacks crest but back of head is squared; dark green above, rusty below, rusty narrow collar, white belly. Female has green breast band. Uncommon; found countrywide at low elevations, in swamps and mangroves and along the wooded edges of streams, marshes, and ponds. Alone or in pairs. Eats small fish and aquatic insects. Digs its burrow nest into embankments or in the dirt clinging to roots of fallen trees; nest is sometimes far from water. The best clue to its presence is often the nervous tick it utters when approached.

---

**Order Piciformes (Woodpeckers and relatives)**

---

**Family GALBULIDAE (Jacamars).** Restricted to the New World tropics; two species in Costa Rica. Jacamars are in the same order as woodpeckers and, like them, have feet with two toes pointing forward, two backward.

## Rufous-tailed Jacamar (*Galbula ruficauda*)

9 in (23 cm). Long slender tail; needlelike bill; glittering green above and on breast, rufous below and on underside of tail; male has white throat, female with buff throat. Uncommon (and inconspicuous); occurs locally on Caribbean and S. Pacific at low and middle elevations. Inhabits understory and middle levels of forest, second growth, and plantations, usually at edges or near clearings. Perches quietly (often in pairs) with bill tilted up as it scans for prey; zips out to grab flying insects, especially butterflies, dragonflies, and bees; beats prey against perch to remove wings. Best means of detection is its call, a sharp loud whistled *peeyu*. Song is an accelerating and ascending series of high-pitched whistled trills.

**Family BUCCONIDAE (Puffbirds).** Puffbirds are restricted to the New World tropics; five species in Costa Rica. Chunky-looking birds with fluffy body plumage, bulky heads, stout bills, and small feet. Sit-and-wait predators principally. Though most species are solitary—and mainly vocalize only at dawn or dusk—the nunbirds and a few other species are both social and vocal. Closely related to woodpeckers.

### White-necked Puffbird (*Notharchus macrorhynchos*)

9.5 in (24 cm). Bold black-and-white plumage; big head; large bill with hooked tip; red eyes. (Pied Puffbird is similar but much smaller; a rare inhabitant of Caribbean lowlands.) Uncommon; occurs countrywide at low elevations. Prefers forests and tall trees at forest edges or in clearings. Perches in middle to upper levels of canopy, on bare horizontal limbs (frequently of large diameter). Flies—in hurtling fashion—on short, rapidly beating wings. Quietly sits for extended periods, waiting for large insects and small lizards to pass within range. Dispatches captured prey (held in bill) by beating it against perch. Places its own nest in a cavity that it carves into arboreal termite nest. Call is a long froglike trill that breaks into stuttering whistles.

**Family RAMPHASTIDAE (Barbets and Toucans).** Barbets occur in tropical regions throughout the world; toucans reside exclusively in the New World tropics. Barbets are named for the bristles at the base of their stout bills. Toucans are characterized by large colorful bills used to grasp fruits and to extract eggs or nestlings from the cavity nests of other birds. Costa Rica is home to two barbet and six toucan species, all resident.

### Prong-billed Barbet (*Semnornis frantzii*)

6.75 in (17 cm). Chunky. Olive above; yellowish head, throat, and breast; red eyes; black mask extends around base of stout gray bill (hook at tip of upper mandible fits into notch at end of lower). Endemic to Costa Rica and western Panama. Common; occurs from Tilarán Mountain Range south to Panama, at middle and high elevations. Inhabits understory of wet forests, second growth, and tall trees in clearings. Sluggish; tends to hop along limbs. Pairs or small groups of birds join flocks of mixed species. Eats mainly fruit, and some insects. Creates its woodpecker-like nest by carving a hole into a dead tree (groups of birds roost together in similar, larger cavities). Pair or flock sings a repeated series of *whah-whah-what* notes (birds not quite in synch, giving the song an echolike quality).

## Emerald Toucanet (*Aulacorhynchus prasinus*)

11.5 in (29 cm). Small. Mostly green with olive crown, blue throat, and chestnut undertail; black and yellow bill is relatively short for a toucan. Common; occurs throughout the country at middle elevations, with seasonal movement to higher and lower elevations. Found at all levels of canopy in forests and in clearings with trees. Flight direct, with fast whirring wingbeats and short glides. Forms small active groups. When foraging, hops and springs through foliage (and often cocks tail, stretches neck, and tilts head in inquisitive pose). Eats copious amounts of fruit but also takes small animal prey, including bird eggs and nestlings. Nests in tree cavity or in hole carved by another species. Song is a monotonous series of *wreck* notes that are at once nasal, raspy, and froglike.

## Collared Aracari (*Pteroglossus torquatus*)

16 in (41 cm). Medium size. Black above; yellow underparts punctuated by dark breast spot and belly band; red rump. Bill is blackish below, ivory above, with dark hash marks along cutting edge. (No range overlap with Fiery-billed Aracari, which occurs in S. Pacific; note orange bill and red belly band on latter.) Common; occurs on Caribbean slope at low and middle elevations (less common at middle elevations of N.W. Pacific). Roams middle and upper levels of forest, tall second growth, forest edges, and semi-open areas with tall trees. In flight, alternates rapid noisy wingbeats with short glides; groups of 5 to 15 fly in follow-the-leader fashion. Eats fruit and small animal prey (and plunders nests of other birds). Up to half a dozen birds cram into the same tree cavity to roost. Call is a high, sharp, squeaky *pe-chees*, repeated irregularly.

## Keel-billed Toucan (*Ramphastos sulfuratus*)

18.5 in (47 cm). Large. Rainbow-colored bill, greenish facial skin, black body with yellow bib, white rump, red undertail, baby-blue legs. Common; occurs on Caribbean slope at low and middle elevations; also in N.W. Pacific (and south to Carara), mostly at middle elevations. Inhabits forest, tall second growth, and open areas with tall trees. Pairs or small groups roam through canopy and below. Long flights display wingbeat bursts alternating with long descending glides. Often perches high up, on snags or exposed branches. Birds in canopy revealed by sound of pattering wingbeats. Eats fruit and occasionally small animals. Nests in large tree cavities. Call is a short croaking rattled *karrrrick*, repeated as bird tosses its bill from side to side.

## Chestnut-mandibled Toucan (*Ramphastos swainsonii*)

22 in (56 cm). Largest CR toucan. Black and yellow bill, chartreuse facial skin, black body with yellow bib, white rump, red undertail, and bluish legs. Common; occurs in Caribbean and S. Pacific at low and middle elevations. Found in forest, tall second growth, and nearby areas with tall trees (strays far from forest much less frequently than does Keel-billed Toucan). Pairs or small groups sometimes encountered sitting quietly, peering down from canopy. Eats fruit and a wide variety of small animals; holds food in tip of bill, tosses it up lightly, then catches it in mouth. Perches on high exposed branch to sing; tilts head back and flips bill as it vocalizes. Far-carrying call is a rhythmic yelping *deeohs-tey-day, tey-day*, often repeated for long periods. Several birds frequently gather to form a chorus at dawn or dusk.

**Family PICIDAE (Woodpeckers).** A large, diverse, cosmopolitan family. Sixteen species in Costa Rica, all resident except the Yellow-bellied Sapsucker, a northern migrant. Sexes alike but for small differences in head plumage. Pecking on hollow trees produces a resonant drumming that conveys territorial and mating messages. Chisel-like bills and barbed tongues are specialized for digging prey out of wood. All have undulating flight.

### Lineated Woodpecker (*Dryocopus lineatus*)

13 in (33 cm). Large. Tufted red crest on male and female (male also has red forehead and moustache); white stripe on face and neck; dark above, fine buff barring below; dark bill. White stripes on back converge but do not meet (cf. Pale-billed Woodpecker). Fairly common; occurs countrywide at low and middle elevations. Mostly in semi-open areas with tall trees and at forest edges (rarely inside extensive forests). Alone or in pairs. Flies high—in undulating fashion—above open ground. Pecks into wood or flakes off bark in search of insects, especially ants; also eats some fruit. Loud vocalizations include a rapid *yak-yak-yak-yak* and a low guttural *chik-gurrrr*. Sound of drum pecking is an accelerating series of sharp taps.

### Pale-billed Woodpecker (*Campephilus guatemalensis*)

14.5 in (37 cm). Large. Bushy red crest and head (female with black forehead and throat); white stripes on side of neck join on back to form "V"; black above, buff barring below; ivory bill. Uncommon; occurs countrywide at low and middle elevations. Pairs inhabit all levels of forest and nearby clearings with tall trees. Chisels deep, oblong, conical holes to get at grubs in rotten trees (standing or fallen). Splinters and chips at the base of a forest tree signal its presence. Vocalization is a nasal bleating. Drum pecking consists of two, quick, powerful taps (usually against a deeply resonant, large-diameter tree).

## Black-cheeked Woodpecker (*Melanerpes pucherani*)

male

7.25 in (18.5 cm). Medium size. Black mask with white patch behind eye; red nape (on male crown also red); fine white barring on back and tail; white rump; grayish underparts (lower half barred black and olive); reddish patch between legs. Common; widespread on Caribbean at low and middle elevations. Inhabits canopy of forest (and lower levels in open areas with tall trees). Alone or in pairs. Hitches up vertical trunks and along limbs, but sometimes also perches briefly on a horizontal twig. Jerks head up and down when agitated. Eats insects, fruit, and nectar. Nests and roosts in holes that it carves into trees. Calls are short churring rattles.

female

## Hoffmann's Woodpecker (*Melanerpes hoffmannii*)

male

7 in (18 cm). Medium size. Yellow nape on male and female (red crown on male); black and white barring above, grayish below; white rump; yellowish patch between legs. Common to abundant; occurs in northern half of Pacific, from low elevations up to mid-elevation valleys and mountain slopes. A recent colonist of deforested areas in northern half of Caribbean. On central Pacific coast hybridizes with Red-crowned Woodpecker (red nape, reddish patch on belly) of S. Pacific. A conspicuous inhabitant of dry forest, wooded areas, plantations, and residential areas with tall trees. Alone or in pairs. Forages at middle and upper levels of canopy, mostly for fruit and nectar but also insects. Carves nest hole into dead trees of small diameter, utility poles, and fence posts. Unmusical vocalizations are dry rattles and chatters.

female

## Chestnut-colored Woodpecker (*Celeus castaneus*)

male

9 in (23 cm). Dark cinnamon with black bars and chevrons; head tawny with a shaggy crest; greenish-yellow bill. Male shows a red patch below eye. Overlaps with similar Cinnamon Woodpecker (smaller, more uniformly rufous, shorter crest). Uncommon; occurs in Caribbean lowlands, and, less commonly, up into foothills. Found in forests and open areas with scattered trees and small woodlots. Alone or in pairs; sometimes accompanies mixed-species flocks. Forages on trunks at middle to upper heights; seems especially fond of trunks covered with dense or tangled vegetation. Eats mainly ants and termites, occasionally fruit. Most often heard vocalization is a single emphatic *peeuu* (husky sounding, somewhat squirrel-like).

female

**Family FURNARIIDAE (Ovenbirds and Woodcreepers).** A large New World family of insectivores. There are 34 species in Costa Rica, all resident. Nearly all are forest dwellers except for the Pale-breasted Spinetail and the Slaty Spinetail, two open-country birds. Coloration and skulking habits make sightings difficult; even when viewed, however, it is often hard to distinguish one species from another. Voice is an important aid to identification. Ovenbirds mostly poke and probe through vegetation; woodcreepers, until recently placed in another family, move along tree trunks in woodpecker-like fashion.

### Northern Barred-Woodcreeper (*Dendrocolaptes sanctithomae*)

11 in (28 cm). Large. Head and body brown, with fine black barring; wings and tail rufous; dark bill is long and stout. Common; occurs countrywide at low and middle elevations (though scarcer in dry N.W. Pacific). Inhabits forests and parklike open areas with trees. Usually alone (less often in pairs). Generally found on trunks, at low to middle heights of medium-sized trees. When attending army-ant swarms, becomes more active and aggressive, sallying to snag fleeing arthropods, small frogs, and lizards. Nests in natural tree cavities. Excited song consists of a series of loud whistled *chu-wee* phrases (inflected down, then up). Among the first birds to sing in the morning; makes inaugural calls well before sunrise. Readily attracted to imitations of its song.

### Spotted Woodcreeper (*Xiphorhynchus erythropygius*)

9 in (23 cm). Medium size. Long, straightish, dusky bill; pale but prominent eye ring; olive-brown with tan spots; rufous wings and tail. Common; occurs countrywide at middle elevations (and less commonly in hills of Caribbean lowlands). Found in moist and wet forests and in older second growth. Alone or in pairs; frequently joins mixed-species flocks. Usually encountered in middle and upper levels of canopy, creeping along trunks and large limbs. Probes moss and epiphytes, flicking aside vegetation in search of insects and small reptiles and amphibians. Like most woodcreepers, the Spotted Woodcreeper peers around trunks to eye observers, then withdraws, moves to a new position, and steals another peak, all the while attempting to conceal body. Song consists of two or three sad whistles (descending and drawn out).

**Streak-headed Woodcreeper** (*Lepidocolaptes souleyetii*)

7.5 in (19 cm). Small and slender. Brownish with buff streaks on underparts, head, neck, and upper back; wings and tail rufous; pale bill is narrow, slightly decurved. (Very similar to several other species, particularly the Spot-crowned Woodcreeper, which replaces Streak-headed in highlands.) Common; occurs countrywide at low and middle elevations (though less numerous in dry N.W. Pacific, where it mostly inhabits mangroves and gallery forest). Seldom found deep within extensive forests, preferring instead patchy forests, forest edges, open woodland, and agricultural land with scattered trees. Generally in loose pairs, less often alone. Active at all levels of canopy, on trunks and branches; probes into bark and flakes it off to get at insects. Seems to forage quickly, spiraling up one tree and then flying down to base of another. Song is a rolling trill that descends in pitch.

**Brown-billed Scythebill** (*Campylorhamphus pusillus*)

9 in (23 cm). Long, thin, deeply decurved bill is diagnostic; body similar in form and coloration to many other woodcreepers. Uncommon; occurs on Caribbean slope at middle elevations, and in S. Pacific both at middle elevations and in lowlands of the Osa Peninsula. Prefers wet forests with dense epiphytic growth. Forages over trunks and branches at all levels of canopy, often in the company of mixed-species flocks. Wields bill with impressive dexterity—forceps-like—to grasp and pry insects from deep within crevices, moss, and epiphytes. Song is varied but often a descending and accelerating series of whistles, the final notes sometimes stuttered. Most vocal at dawn and dusk.

### Wedge-billed Woodcreeper (*Glyphorhynchus spirurus*)

6 in (15 cm). Only CR woodcreeper with a short, wedge-shaped bill. (Plain Xenops and Streaked Xenops, of same family, have short tails and white facial stripes; they forage acrobatically on small branches.) Common; occurs on Caribbean and S. Pacific at low and middle elevations. Inhabits forest and woodland; also ventures into nearby open areas with clustered trees. Usually alone. Spirals up tree trunks, flaking off bits of bark to uncover tiny insect prey, then flying down to the base of another tree. Tends to stay rather low on tree and seems to prefer large trees. Places nest in a narrow cavity formed between closely spaced root buttresses. Often emits its sharp, flicking call; soft slurred trilling song is seldom heard except at dawn.

### Red-faced Spinetail (*Cranioleuca erythrops*)

6 in (15 cm). Pointy graduated tail; narrow bill; olive-brown body; rufous wings, tail, and front half of head. Common; occurs countrywide at middle elevations (and locally higher). Inhabits middle levels of wet epiphyte-laden forests. Alone or in pairs; often joins mixed-species flocks in which Common Bush-Tanagers are the lead birds. Actively creeps and hops over limbs and foliage. Twists and turns when rummaging through tangled vegetation, epiphytes, and dead-leaf clusters. Nest (entered from the bottom) is a conspicuous volleyball-sized clump of moss dangling at the end of a branch, aerial root, or vine; it is usually placed within a clearing or over a trail or road. Song is an accelerating twitter of high-pitched lisping notes.

**Family THAMNOPHILIDAE (Antbirds).** Members of this family occur exclusively in the New World tropics. There are 22 species in Costa Rica, all resident (and all insectivorous). Some antbird species have the habit of following army-ant swarms to catch the insects that flee the swarm, thus the common name. Most have a concealed patch of pale feathers (on back, wings, or crown) that is exposed when a bird is agitated or during courtship.

## Barred Antshrike (*Thamnophilus doliatus*)

6.25 in (16 cm). Erectile crest; stout bill with hooked tip; pale-yellow eyes. Male with black-and-white barring; female (and young) rufous above, buff below, with black streaks on face and nape. (Male similar to male Fasciated Antshrike of Caribbean lowlands.) Fairly common; occurs countrywide at low and middle elevations. Found at forest edges and in scrubby woodlands and second-growth thickets; in drier regions, prefers evergreen forests. Pairs are fairly conspicuous, despite moving low through thick cover. When scanning for small insects, hops deliberately, then gleans prey from vegetation. Nest is a cup of woven fibers that hangs between a forked twig. Song is an accelerating series of nasal *wah* notes that descends in pitch and ends with a sudden, barked *wenk*.

male

female

## Black-hooded Antshrike (*Thamnophilus bridgesi*)

6.5 in (16.5 cm). Fairly stout bill with hooked tip; small white spots on wing coverts. Male black with gray belly; female brownish with white streaks on head and underparts. Common; occurs in S. Pacific at low and middle elevations. Endemic to Costa Rica and western Panama. Inhabits forest edges, clearings, and tall tangled second growth (especially where there are dense vines). Occurs in pairs that sometimes join mixed-species flocks or wander in the proximity of the flocks. Perches at middle levels of canopy; peers about, then hops and flits toward insect prey. Often allows observers a close approach. Song is a slowly accelerating series of same-pitched notes that ends with a final drawn-out note; pounds tail downward to the beat of its song

male

female

## Chestnut-backed Antbird (*Myrmeciza exsul*)

5.5 in (14 cm). Chunky; short tail; chestnut body; sky-blue patch of bare skin around eyes; black on head is extensive on male, limited on female. Common; occurs on Caribbean slope and S. Pacific in lowlands and foothills. Inhabits understory of wet forests, tall second growth, and tangled thickets along forest edge. Almost always in pairs; sometimes attends army-ant swarms or associates loosely with mixed-species flocks. Rummages through dense undergrowth in search of arthropod prey, hopping on ground and occasionally clinging to vertical stems. Incessantly pounds tail. Places untidy cup nest at low point in dense vegetation. Song consists of two or three emphatic whistles: *cheap, cheap, cheeeer.* Call is a jeering nasal chatter.

female

male

**Family FORMICARIIDAE (Antthrushes).** Found only in New World tropics. Three species in Costa Rica, all resident. Plump birds with long legs, short cocked tails; in gait and appearance somewhat chickenlike. Insectivorous. Most stalk the floor of mature forests and are rarely seen.

## Black-faced Antthrush (*Formicarius analis*)

7 in (18 cm). Plump body; short cocked tail; brown above, gray below; black throat and face; rufous on side of neck; bare bluish skin around eye imparts big-eyed look. Sexes alike. Uncommon; occurs in S. Pacific in mature wet forests of lowlands and middle elevations; also occurs on Caribbean slope in lowlands (replaced by Black-headed Antthrush at middle elevations). Walks over shaded forest floor, pausing to flick aside leaf litter in search of insects. Nest, a natural chimneylike cavity in stump or snag, is entered from the top. Typical song consists of three mellow whistled notes (*keep, two, two*) that are sometimes run into a longer series. When flushed, utters one or two sharp whistles in alarm.

**Family TYRANNIDAE (Tyrant-Flycatchers).** An exclusively New World family. There are 80 species in Costa Rica (more than any other family); of these, 20 are migrants, either visitors from the north or birds that breed here then move south. Sexes and ages usually alike. Extremely difficult to distinguish one species from another; best means of identification is often voice. Most have a concealed crown patch of colorful feathers that are exposed during displays. Song at dawn often distinct from calls made during the day. Many species have long, stiff bristles around the bill that aid in catching flying insects.

## Yellow-bellied Elaenia (*Elaenia flavogaster*)

6 in (15 cm). Spiky crest, white crown patch; olive-brown overall except for yellowish belly and pale wing bars; small bill; long tail; erect posture. (Overlaps in S. Pacific with very similar Lesser Elaenia.) Common; occurs countrywide at low and middle elevations (and locally higher); less numerous in dry N.W. Pacific. Usually in pairs. Inhabits open areas with scattered trees (such as yards and gardens), scrubby woodland, borders of pastures, and agricultural plots. Found at middle and upper levels of canopy, actively moving through vegetation to glean insects and pluck berries; may sally short distances to catch flying insects. Noisy and excitable; often sits atop trees or shrubs to vocalize. Call is a burry, mournful whistle. Pairs engage in a raucous and urgent duet of harsh stuttering whistles.

crest raised

## Mountain Elaenia (*Elaenia frantzii*)

6 in (15 cm). Rounded head lacks crest; olive to olive-brown overall, paler below; pale eye ring; wings show two pale bars and pale feather edgings. Posture more horizontal than in other *Elaenia* species. Abundant when breeding (Feb. to Aug.); occurs countrywide at high elevations. After breeding many birds descend to foothills (or lower), and some perhaps migrate south out of CR. Inhabits canopy of tall forest, forest edge, and second growth (also hedgerows and shrubs in more open areas). Generally alone or in pairs, though sometimes congregates at fruiting trees. Eats mostly berries but also gleans insects from vegetation or catches them in the air. Tirelessly repeats call, a scratchy drawn-out whistle, from treetops of highland forest.

## Torrent Tyrannulet (*Serpophaga cinerea*)

4 in (10 cm). Tiny. Gray overall, whitish below; black head, wings, and tail. Common; occurs in S. Pacific and on Caribbean slope of the Tilarán Mountains, in foothills and higher (absent from Guanacaste Mountains). Usually seen in pairs. Flits along shores of rushing mountain streams and perches atop boulders in riverbed. Can be quite tame. Catches small flying insects, also snatches aquatic insects and larvae from water's edge. Bobs tail up and down. Makes a sharp *chip*; pair members excitedly repeat *chip* call when reuniting.

## Paltry Tyrannulet (*Zimmerius vilissimus*)

3.75 in (9.5 cm). Tiny. Grayish green; pale line above eye (eyes pale too); wing feathers with yellow edge (no wing bars); stubby bill. Common and widespread from lowlands to highlands, but most common at low and middle elevations, in wetter environments. Inhabits forest canopy, forest edge, woody second growth, and scattered trees in open areas. Usually seen alone, sometimes in pairs. Forages actively at middle to upper levels of trees; eats primarily mistletoe berries (Mistletoe Tyrannulet is alternate name). Regurgitates sticky mistletoe seeds, and is often seen wiping bill on branch to free itself of these. Call is a plaintive *peeyup*.

## Common Tody-Flycatcher (*Todirostrum cinereum*)

3.75 in (9.5 cm). Tiny. Blackish gray above, yellow below; white tail tips; wings have yellow edges; yellow eyes; bill is long, broad, and flat. Long graduated tail is held cocked. Common; occurs countrywide at low and middle elevations. Prefers forest edge, second growth, and open areas with trees and shrubs. Pairs forage at eye level or above, making short sallies to grab insects off foliage (often with an audible snap of the bill). Slender nest—usually placed in the open—hangs from branch tip; nest looks like a long pointy beard and has a shielded side entrance. Calls include a repeated insect-like ticking and a buzzing trill.

**Tufted Flycatcher** (*Mitrephanes phaeocercus*)

4.75 in (12 cm). Small. Elegant pointed crest; upright posture; olive-brown above, bright cinnamon below (shading to yellow on belly); pale eye ring. Fairly common; occurs from highlands down to wet middle elevations. Pairs inhabit clearings within tall forest; less often found in nearby open areas with scattered trees. Frequently encountered near streams. From an exposed perch, scans for flying insects while jerking head about; sometimes engages in lengthy flights to catch prey; after settling back on perch, often briefly wiggles its tail. Whistled call is a quick ringing *peet, peet, peet* (sometimes repeated in longer sequences).

**Black-capped Flycatcher** (*Empidonax atriceps*)

4.5 in (11.5 cm). Small. Olive-brown overall; whitish throat, faint pale wing bars, sooty head; white eye ring interrupted above eye. (There are eight *Empidonax* species in CR; the majority are migrants and nearly all notoriously difficult to identify. This species, however, is a CR resident and readily identifiable.) Endemic to Costa Rica and western Panama. Common; occurs in central and southern mountain ranges at high elevations. Usually alone, perched on low vegetation in shrubby clearings and at forest edges (also occurs in canopy of tall forest). Habitually flicks tail. Quite tame. Catches small flying insects, often returning to the same perch after sallies. Call is a simple *whit*.

**Long-tailed Tyrant** (*Colonia colonus*)

5 in (13 cm). Two central tail feathers project well beyond tail tip. Pale-gray cap bordered by white stripe running through eye. Grayish-white stripe down its back. Fairly common; occurs on Caribbean slope at low elevations and in foothills. Found mostly along forest edges and in clearings with tall trees. Usually in pairs, on high, exposed perches. Eats flying insects caught in elaborate aerial pursuits. Nests in old woodpecker hole or small natural tree cavity. Utters a high thin *pweep* (a bit drawn out). Pairs frequently call back and forth.

## Dusky-capped Flycatcher (*Myiarchus tuberculifer*)

6.5 in (16.5 cm). Blackish cap, slightly peaked crest. Upperparts are olive-brown; pale margins on wing feathers; gray on throat and breast blends to yellow on belly. (Of the six *Myiarchus* found in Costa Rica—all quite similar—this is the only species with a dark cap.) Common; occurs countrywide except in high mountains. Inhabits all levels of forests, forest edges, and open woodland. Usually solitary. Snatches insects off the ground and off vegetation, also eats many berries. Nests in an old woodpecker hole or similar cavity in a small tree or fence post; like most *Myiarchus*, adorns nest with shed snake skins. Most often heard vocalization is a mournful, descending whistle.

## Great Kiskadee (*Pitangus sulphuratus*)

9 in (23 cm). Bright-yellow breast; bold stripes along head meet on nape; rufous on wings and tail. (One of six very similar birds in Costa Rica; so similar that local Spanish common name, *pecho amarillo*, is applied to all six. Most similar to this bird is the Boat-billed Flycatcher, with a broader bill and olive on wings and tail. Also compare with Social Flycatcher.) A common and widespread resident; absent only from high mountains. Favors open areas, especially those near water. Usually in pairs or boisterous family groups. Quite omnivorous for a flycatcher; eats fruit, invertebrates, and small vertebrates (including fish). Nest is a loose dangling ball of fibers and straw; placed in trees, cactus, or manmade structures (often seen woven between wires on utility pole). Very vocal; name derived from loud *kis-ka-dee* song.

## Social Flycatcher (*Myiozetetes similis*)

6.25 in (16 cm). Olive above; dark cheeks; white eye stripes continue behind eye but do not join at nape; red crown patch, normally concealed, is exposed when bird becomes agitated. Common; occurs countrywide at low and middle elevations (absent from high mountains). Found in scrubby areas and open clearings with scattered trees, often near human habitations. Found in pairs. Usually perches high. Roofed nest is often usurped by Piratic Flycatcher. Catches insects and eats berries. Call is a repeated emphatic *FEEur*.

## Tropical Kingbird (*Tyrannus melancholicus*)

7.5 in (19 cm). Grayish olive above; pale yellow below. Grayish head slightly peaked, subtle dark mask, whitish throat; notched tail. Abundant countrywide, becoming scarce only at high elevations or in extensive forests. From an exposed perch, hunts flying insects with extreme agility and persistence; also feeds on berries. It is a bold and relentless pursuer of large predatory birds that intrude upon its territory. Usually in pairs but also gathers in communal roosts when not breeding. Often utters high-pitched excited tittering calls. One of the easiest birds to see in Costa Rica, as it is abundant and perches on roadside electrical wires.

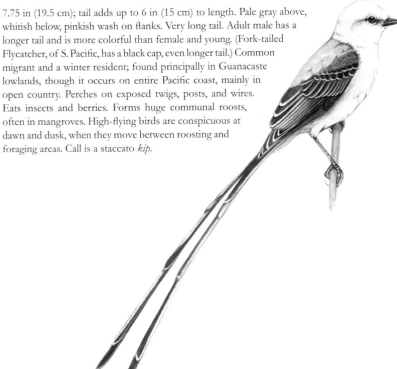

## Scissor-tailed Flycatcher (*Tyrannus forficatus*)

7.75 in (19.5 cm); tail adds up to 6 in (15 cm) to length. Pale gray above, whitish below, pinkish wash on flanks. Very long tail. Adult male has a longer tail and is more colorful than female and young. (Fork-tailed Flycatcher, of S. Pacific, has a black cap, even longer tail.) Common migrant and a winter resident; found principally in Guanacaste lowlands, though it occurs on entire Pacific coast, mainly in open country. Perches on exposed twigs, posts, and wires. Eats insects and berries. Forms huge communal roosts, often in mangroves. High-flying birds are conspicuous at dawn and dusk, when they move between roosting and foraging areas. Call is a staccato *kip*.

**GENERA INCERTAE SEDIS (Becards and relatives).** This group is composed of 11 species that appear to be closely related to the Tyrant Flycatchers. Future genetic work will determine their true taxonomic affinity.

## Rufous Piha (*Lipaugus unirufus*)

9 in (23 cm). Plumage entirely rufous. Ages and sexes similar. (Best distinguished from almost identical but smaller Rufous Mourner by vocalizations.) Uncommon; occurs on Caribbean and Pacific slopes at low and middle elevations. Generally alone, less often in loose groups. Found in forest interior at middle and upper levels of canopy. Sits quietly for extended periods, then flies out to snatch fruit or large insect with its bill. More often heard than seen, its calls are explosive whistles and chatters given at irregular intervals; utters a whistled *weeo-weet* (like someone hailing a taxi).

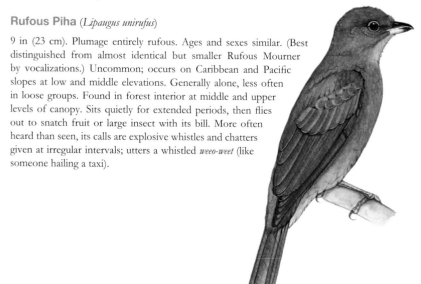

## Masked Tityra (*Tityra semifasciata*)

8.3 in (21 cm). Robust; large head; short tail; striking pink on face and bill. Male white; black sub-terminal tail band; black forecrown and mask. Female and young similar to male but dull gray above, no black on head. Common; occurs countrywide at low and middle elevations. Found in forest canopy and in open areas with tall trees. In pairs or small groups. Often perches conspicuously in treetops. Eats mostly fruit, especially figs, but also takes large insects and small lizards. Nests in tree cavities and holes; pairs may commandeer active woodpecker nests. One Spanish common name, *pájaro chancho* (pig bird), refers to its nasal, *oink*-like calls.

male

female

**Family COTINGIDAE (Cotingas).** A very diverse family with many unresolved taxonomic issues. Confined to the New World tropics; seven species in Costa Rica. Males are often brilliantly colored and generally larger than females. Females responsible for all nesting duties. Most species are strictly frugivorous.

## Turquoise Cotinga (*Cotinga ridgwayi*)

7 in (17.5 cm). Male turquoise blue with deep-purple patches on throat and belly. Female grayish buff above, whitish below, with dark spots overall. (The Lovely Cotinga is nearly identical but occurs only on Caribbean slope.) Uncommon; endemic to Costa Rica (in S. Pacific) and Panama. Frequents canopy of mature forest; also found in tall trees in open country. Generally seen alone but may congregate to feed at fruiting trees and for courtship. Especially fond of figs and the olivelike berries of *Psittacanthus* mistletoes. In flight, males produce a tinkling rattle with wings, and in courtship emit a sharp high-pitched whistle; otherwise, generally silent.

male

female

## Purple-throated Fruitcrow (*Querula purpurata*)

11 in (28 cm). Black overall; crowlike but with broad wings and short tail. Adult male has a purple throat-patch (sometimes extended in display). Fairly common; occurs on Caribbean at low and middle elevations. Inhabits upper levels of forests and forest edges. Small, noisy groups of up to 8 birds swoop through forest, often mixing with nunbirds, toucans, oropendolas, and caciques. Fruit diet supplemented with occasional large insects. Flocks easily located by constant banter of rich, querulous whistles and occasional harsh, raspy squawks. Nest is a flimsy platform of sticks.

female

male

## Three-wattled Bellbird (*Procnias tricarunculatus*)

Male 12 in (30 cm); female 10 in (25 cm). Adult male is chestnut with a pure-white hood; three noodlelike dark wattles droop from the base of the bill. Female olive above, yellowish below; dark-olive streaks overall. Femalelike plumage on immature male gradually changes to adult plumage over a 5-year period. Uncommon; breeds in highlands of major mountain ranges, then descends to foothills and lowlands, where it can be found in almost any forested region (though visits are often brief). Mostly solitary outside of breeding areas. Frequents upper levels of forest. Feeds on large fruits; especially fond of wild avocados. Males form leks in canopy to call and display to females. Male's lek song consists of a few very loud explosive *dong!* notes followed by softer, thin whistles and bubbling chatter. Regional dialects are distinguishable.

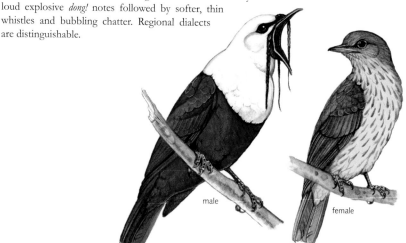

male

female

# Lekking

Three-wattled Bellbird (*Procnias tricarunculatus*).

Some birds form leks, an assemblage of males at a traditional site in which the males vie for the attention of females through competitive vocal displays and choreographed dance performances. Leks form every day during the breeding season, with each male establishing himself at a designated spot within the lek. In these so-called classical leks, males spar with their neighbors for mating rights.

Exploded leks are an entirely different kind of affair. In these leks, males may be spaced several hundred yards apart, and rely on vocal signals rather than physical displays to initially attract females. The male Three-wattled Bellbird, for example, issues a loud *dong!* that resounds throughout the forest, attracting distant females. The bellbird learns its song during development. Over time, the birds within a given area modify their song to create a dialect, an indication of the importance of song to mating success. When a female bellbird approaches a singing male, he often performs a "changing-place" display. The male jumps toward the hen, who simultaneously moves to the spot just vacated by the male. He then sidles over to the female and calls directly at her. If she is suitably impressed, mating follows.

**Family CORVIDAE (Crows and Jays).** A cosmopolitan family of birds noted for their intelligence. Sexes are alike. Most species have large bodies; all have straight stout bills and strong feet. Many exhibit social breeding systems and commonly form flocks. The only corvids in Costa Rica are jays, of which there are five species, all resident.

## White-throated Magpie-Jay (*Calocitta formosa*)

18 in (46 cm). Ranges from sky blue to turquoise above; white below; conspicuous crest of curved feathers; long, graduated tail. Common in dry lowlands and foothills of N.W. Pacific; uncommon on central Pacific coast. In pairs or small loose flocks. Flocks frequently mob predators. Favors dry forest, scrub woodland, open country with scattered trees, and farms. Flight deliberate but graceful (long tail streams behind). Noisy and curious. Birds hop on the ground and among trees to procure a wide variety of plants and animals. A cooperative breeder, with several adults tending one nest. Make a wide variety of calls, some soft and mellow, others loud, harsh, even mechanical sounding. *Illustration not to scale.*

## Brown Jay (*Cyanocorax morio*)

15.5 in (39 cm). Brownish hood, wings, and back; creamy white below. Long, dark tail is graduated, with broad white tip. Sexes alike but young have yellow eye ring and yellow bill (brownish on adults). Common; occurs in deforested areas, in lowlands on Caribbean slope and in mountains and foothills in the rest of the country (generally absent in southern Pacific region though it is expanding its range southward). Prefers open country with scattered trees, sparse woodland, and forest edges. Birds actively search at all levels for invertebrates, small vertebrates, fruits, nuts, and nectar. In flight, deep, springy wingbeats alternate with short glides. An irritable and noisy species; forms flocks of up to a dozen birds that are quick to sound off at the slightest suggestion of threat. Local name *piapia* derives from raucous *piyaaah* calls. *Illustration not to scale.*

adult

young

**Family PIPRIDAE (Manakins).** Forest birds of the New World tropics; eight species in Costa Rica, all resident. Small, stocky birds with heads that are large relative to the body. Frugivorous. Males are usually black with bold, brightly colored patches. Females are drab, often greenish. Males perform ritualized courtship displays. Females take care of all nesting duties.

## White-collared Manakin (*Manacus candei*)

4.25 in (11 cm). Flat-headed; orange legs. Male has black cap, broad white collar, black back and wings, and yellow belly. Female and young are drab green overall, yellowish below; nearly identical to female and young Orange-collared Manakin, of S. Pacific (endemic to Costa Rica and Panama). Common; occurs throughout Caribbean lowlands and foothills. Found alone or in loose groups. Inhabits low, thick vegetation along forest margins of mature and secondary forest. Eats berries of understory shrubs and trees, and also insects. Male leks are situated in dense stands of saplings; each male clears a small patch of bare ground at the base of a vertical stem and then hops back and forth between ground and stem. Local name is *quiebra ramas* (twig breaker) in reference to popping sound made by males' wings during lek displays.

male

female/
young

young male

## Long-tailed Manakin (*Chiroxiphia linearis*)

4.5 in (11.5 cm); male's tail adds up to 6 in (15 cm) to length. Both sexes have orange legs. Adult male is black with red crown, sky-blue back, and two narrow, elongated central tail feathers. Female and young are grayish olive; young male takes several years to reach full adult plumage. Common; occurs on northern half of Pacific slope, including the Central Valley, from lowlands to middle elevations. Found alone or in small groups. Active at low and middle levels of forest, especially in vine tangles. Two males pair up (usually a young male and an old male) to attract females by performing synchronized vocalizations and a courtship dance in which each male rapidly leapfrogs over the other along a branch. Quite vocal; produces short, mellow whistles, and nasal meows. Costa Rican name, *toledo*, is a mnemonic for male's *toe-lay-doe* lek call.

male

female

## Red-capped Manakin (*Pipra mentalis*)

4 in (10 cm). Adult male is black with a brilliant-red head, yellow thighs, and white eyes. Female and young are dull olive. Common; occurs on Caribbean and S. Pacific in lowlands and foothills. Found in the interior of wet forests, at middle and upper levels. Away from courtship arenas it is solitary. Eats berries and snatches insects in short sallies. Male's courtship dance is performed on a relatively high limb. He stretches out his legs to expose yellow thighs, engaging in a variety of antics that include quick stutter steps along the limb, abrupt about-faces, and repeated short flights of looping and arcing patterns. All this is accompanied by bizarre, typewriter-like wing snapping and sharply whistled vocalizations.

male

female

**Family HIRUNDINIDAE (Martins and Swallows).** Martins are larger and bulkier than swallows, with proportionally larger heads and with broader bills. Endowed with long pointed wings, these aerial specialists catch insects in flight. Gregarious. Thirteen species in Costa Rica, three resident, eight migrant, and two species with both resident and migrant populations.

### Gray-breasted Martin (*Progne chalybea*)

6.5 in (16.5 cm). Male steely blue above; gray breast and flanks, white belly. Female shows more gray. Young are brownish. Common; occurs locally throughout the country from low elevations to middle elevations (and above). Abundant in many cities, where it perches on wires and building ledges and nests in holes in buildings (especially fond of metal structures at gas stations). Outside of cities, it occurs in open country, often near large rivers; nests in tree cavities and old woodpecker holes. In flight, frequently soars and glides; often forms large flocks. Makes husky *chirp* calls.

### Mangrove Swallow (*Tachycineta albilinea*)

5 in (13 cm). Adults glossy-green or blue above; white below; prominent white rump; narrow white "V" on forehead. Young grayish brown above; smudgy wash on sides of chest. Common; occurs on Caribbean and Pacific slopes at low elevations and, locally, at middle elevations. On coasts it inhabits estuaries, lagoons, and salt and aquaculture ponds; inland it can be seen along large slow-moving rivers and above lakes, marshes, and wet pastures. Skims low over water. Tends to perch on fairly low branches or wires. Typically occurs in pairs or small flocks; more gregarious when not breeding (in late rainy and early dry season). Places nest in a crevice, usually near water.

## Blue-and-white Swallow (*Pygochelidon cyanoleuca*)

4.25 in (11 cm). Adults are a metallic blue-black above; white below except for dark undertail, which is unique among Costa Rican swallows. Young show a brownish tint above; buff colored below. All ages have a forked tail. Common; local populations occur countrywide at middle and high elevations, though scarcer in the Guanacaste Mountain Range and absent from Nicoya Peninsula. Migrants from South America (present between May and Sept.) may turn up anywhere in the country. Found in pairs or small flocks. Flocks circle and dive in relatively slow flight, in and around towns, rural homes, agricultural areas, and forest clearings. Places nest in crevices and on ledges, on both buildings and rock faces. Birds call frequently, producing sweet chirps or a scratchy, high-pitched chatter.

## Southern Rough-winged Swallow (*Stelgidopteryx ruficollis*)

4.75 in (12 cm). Brownish overall; cinnamon wash on throat; pale rump; tail has slight notch. Age and sex differences are slight. (Range overlaps with that of Northern Rough-wing, which lacks cinnamon throat and pale rump.) Common; occurs on both Caribbean and Pacific slopes in humid lowlands (and higher); absent from dry N.W. Pacific. Found in open country; often burrows its nest into earthen embankments running along creeks and road cuts. In pairs or small flocks. Perches on wires or on leafless branches; flies directly, with little gliding, usually staying low but sometimes flying far above ground. Call is a short, rolling *chirp*.

## Barn Swallow (*Hirundo rustica*)

7.5 in (19 cm). Long, deeply forked tail shows narrow band of white spots (only swallow in Costa Rica with white on tail). Glossy blue-black above. Adult male with chestnut forehead and orangish breast; adult female paler below; young duller overall. Birds in molt have short tails and faded or splotchy plumage. Common to abundant winter resident, arriving in Aug. and departing in May; principally found at low elevations, in open agricultural areas, pastures, and marshes. Migrating birds occur countrywide, but are especially numerous along both coasts. Usually in flocks that skim low over vegetation to catch flying insects. Perches in compact groups on electrical wires, fences, low vegetation, or directly on the ground. Forms large communal roosts.

**Family TROGLODYTIDAE (Wrens).** There are 22 species in Costa Rica, all resident. Many have barring on wings and/or tail. There is little or no difference in plumage between ages and sexes. Songs are often melodious, and in many species pairs sing duets. Insectivorous. Wrens sleep alone or in family groups, either inside existing cavities or in the nestlike dormitories that they build.

## Band-backed Wren (*Campylorhynchus zonaris*)

6.5 in (16.5 cm). Large. Dark and buff barring above; whitish below with heavy dark spotting; cinnamon wash on lower section of underparts; long tail. Common; occurs on Caribbean at low and middle elevations. Favors broken forest, forested riversides, and plantations and yards with trees. Highly social; forms noisy family groups of up to a dozen birds. Mostly at upper levels, only rarely descending to the ground. Scrambles over vegetation, poking and probing into epiphytes and under bark in search of insects. A cooperative breeder; several adults attend nest and young. Calls constantly, emitting an emphatic, dry chattering. When several birds call together the sound suggests a dog fight more than a song.

## Rufous-naped Wren (*Campylorhynchus rufinucha*)

6.75 in (17 cm). Large. Bold head pattern; rufous nape and back; white underparts; long, white-tipped tail. Common; occurs in N.W. Pacific at low and middle elevations; also in Central Valley. Frequents yards and gardens, dry and evergreen forests, scrub woodland, and agricultural areas with trees. Conspicuous and vocal. In pairs or small family groups. Tends to move at low and middle levels in search of insects, poking into vegetation or nooks and crannies of buildings. Nest is a big ball of loose fibers often placed in a palm, cactus, or Bull-horn Acacia and also woven into wires of a utility pole; entrance to nest is on side. Birds maintain a constant raspy chatter and sing duets of rich bubbling whistles.

**Bay Wren** (*Thryothorus nigricapillus*)

5.75 in (14.5 cm). Bright chestnut body; white throat; black head with white facial markings. Common on Caribbean slope at low and middle elevations. Found in forests, generally in thickets near streams. In pairs or small family groups. Explosively loud song is composed of rich slurred whistles; call notes are staccato, harsh, and grating. (There are 10 members of the genus *Thryothorus* in Costa Rica. They tend to inhabit low, dense vegetation, and are more often heard than seen. Pairs stay together throughout the year and sing melodious antiphonal duets.)

**House Wren** (*Troglodytes aedon*)

4 in (10 cm). Small. Brown with fine indistinct barring on wings and tail. Common to abundant; occurs virtually countrywide, though generally absent in dry N.W. Pacific. Avoids deep forest in favor of open woodlands, agricultural areas, roadsides, and gardens. Usually solitary. This weak flyer prefers to hop or scurry on ground and through low vegetation. Active and curious. Enters buildings and pokes into crevices and holes in search of insects. When foraging, maintains a soft, garrulous chatter; rich and bubbly song is composed of varied warbles and trills.

**Gray-breasted Wood-Wren** (*Henicorhina leucophrys*)

4.25 in (11 cm). Small and chunky. Chestnut above, gray below; dark cap, white eye line; stubby, erect tail. Common; occurs in wet forests on all the major mountain ranges, from middle elevations up to tree line. It is replaced by the White-breasted Wood-Wren in lowlands (ranges do overlap at middle elevations). Vocal pairs scurry rodentlike through understory vegetation of dense forests, sometimes risking a brief peek at human observers. Individuals scold with hard dry rattling calls; pairs sing a rich and rollicking duet that is a pervasive sound of highland forests.

**Family TURDIDAE (Robins and Thrushes).** There are 15 species in Costa Rica; 11 are residents, 4 are migrants from North America. Most have plumages of subdued colors. Many species produce beautiful songs. Generally sexes are alike; young are duller than adults and heavily spotted.

## Black-faced Solitaire (*Myadestes melanops*)

6.75 in (17 cm). Adult slate gray; black mask; orange bill and legs. Upright posture. Uncommon in most of its range, but can be locally common; occurs at middle and high elevations of major mountain ranges. On Caribbean slope, it sometimes moves to lower elevations between Aug. and Feb. Inhabits wet forests; less frequently in broken forest or tall second growth. Usually alone. Tends to remain at low levels of the canopy, though it will venture higher in search of fruits and berries. Its hauntingly beautiful song is composed of long metallic trills and quavering whistles. It is a highly prized cage bird that has been extirpated or reduced in number in many areas.

young

adult

## Black-billed Nightingale-Thrush (*Catharus gracilirostris*)

5.75 in (14.5 cm). Fairly small. Olive-brown overall with gray throat and belly; black bill. Very common; occurs in the Central and Talamanca mountain ranges at high elevations. Endemic to Costa Rica and Panama. Found in forests and nearby clearings. Alone or in pairs. Hops on ground and through low vegetation. Has an erect stance; it nervously flicks tail up as it lowers its wings. Eats berries and insects. Confiding; often allows close approach. Flutelike song consists of several muted, tinkling whistles. (Costa Rica has five nightingale-thrush species. Each occupies a unique elevation niche; ascending mountain slopes, one species gives way to another, with little overlap.)

### Sooty Thrush (*Turdus nigrescens*)

10 in (25.5 cm). Large. Male is black, female brownish; in both sexes, eyes white (with orange eye ring); bill and legs are orange. Common; occurs in the Central and Talamanca mountain ranges at high elevations (most common above tree line). Endemic to Costa Rica and Panama. Alone or in loose groups. Hops along open ground in search of insects; also snatches berries from trees and shrubs. Call consists of harsh rolling notes; unlike most robins, its song is fairly weak and unmusical.

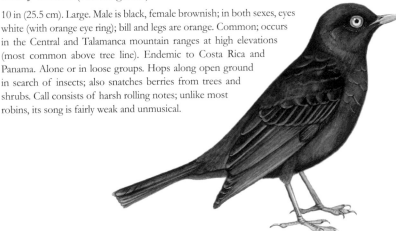

### Clay-colored Thrush (*Turdus grayi*)

9.25 in (23.5 cm). Brown overall; yellowish bill; reddish-brown eyes. Sexes alike. Common; occurs throughout the country except at the highest elevations; mostly in open habitats. Pokes through leaf litter in search of worms and other small invertebrates; also eats a variety of fruits and is a regular visitor at banana feeders. Places its mud and fiber cuplike nest on buildings as well as in trees. Local name is *yigüirro*. This is the national bird of Costa Rica, selected not for its drab plumage but because of its beautiful song. According to local folklore, this bird sings at the end of the dry season to call forth the rains.

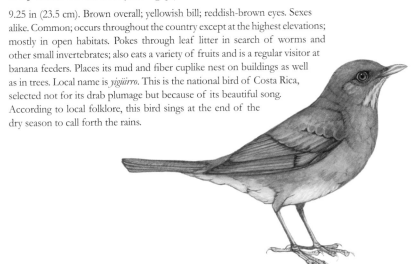

**Family PTILOGONATIDAE (Silky-Flycatchers).** The four members of this small family were formerly lumped into the Waxwing family (Bombycillidae). The two species that occur in Costa Rica are endemic to Costa Rica and Panama.

## Long-tailed Silky-Flycatcher (*Ptilogonys caudatus*)

9.25 in (24 cm). A slim, handsome bird. Gray overall with olive-yellow on head and underparts; bushy crest; long graduated tail terminates in a spike; small bill, short legs. Female more olivaceous; tail is shorter. Common; occurs in Central and Talamanca mountain ranges at high elevations. Found in forest and open, disturbed areas. Catches flying insects and plucks berries from trees and bushes. Pairs or small flocks often seen perched conspicuously atop tall trees or flying high overhead. In flight, frequently makes tinkling or metallic, sputtering calls.

**Family PARULIDAE (Wood Warblers).** A large New World family of spritely, colorful birds. There are 53 species that occur in Costa Rica; 39 are northern migrants, 13 are residents, and 1 has both resident and migrant populations. Many species are very similar in appearance; sexual dimorphism and plumage variations pose further challenges for correct identification. Migrant species rarely sing when in the tropics.

## Tennessee Warbler (*Vermivora peregrina*)

4.25 in (11 cm). Generally dull greenish above, paler below; white undertail; no wing bars. Narrow dark line through eye is bordered above by pale line. Males in breeding plumage show a gray head. (Note that yellow or orange smudges on head and face may be pollen stains, not true markings.) Common to abundant winter resident, arriving in late Sept. and departing by the end of April; occurs countrywide except at highest elevations. Inhabits treetops in forests, plantations, yards, and gardens. In Costa Rica it supplements its insectivorous diet with berries and flower nectar. Often found in small single-species flocks, though it also mixes with other birds at fruiting and flowering trees.

breeding
male

## Yellow Warbler (*Dendroica petechia*)

4.5 in (11.5 cm). Yellow overall; olive on wings and tail; no wing bars. Adult male bright yellow with chestnut streaking on breast and flanks; adult female pale yellow with little or no chestnut streaking; young birds, especially females, can be very drab. Resident adult male has rufous head and neck; resident adult female sometimes has a rufous crown. Resident population restricted to mangroves on both coasts. Abundant migrant and winter resident, arriving in early Sept. and departing in early May; occurs countrywide at low and middle elevations. Prefers open parks and gardens, plantations and agricultural lands, low scrub and dry forest. Winter residents defend a territory and are solitary, gleaning insects from vegetation while frequently uttering a sharp *chip*. Song (heard only in mangroves from resident birds) resembles *sweet-sweet-sweet-I'm so sweet*.

resident
adult male

migrant
adult male

young female
(both races)

## Chestnut-sided Warbler (*Dendroica pensylvanica*)

4.5 in (11.5 cm). Ages and sexes differ but birds of all plumages show pronounced wing bars. When standing, cocks tail up and droops wings slightly. Breeding female resembles breeding male but has less chestnut on flanks (and markings on face are less distinct). Nonbreeding adults and young birds are yellow-green above, pale below; gray face; white eye ring; chestnut on flanks reduced or absent. An abundant migrant and winter resident, arriving in Sept. and departing in April; occurs countrywide at low and middle elevations, though less numerous in dry N.W. Pacific. Prefers forest and woodland, but also occurs in open areas with scattered trees. Hops and flits actively through foliage at middle and upper levels of trees and tall shrubs, gleaning insects from vegetation. A dull *chip* note is the only vocalization it emits when in Costa Rica.

breeding male

young female

## Northern Waterthrush (*Seiurus noveboracensis*)

5.25 in (13.5 cm). Brown above; underparts with dark streaking and pale-yellow wash; buff or yellowish line above eye. (Louisiana Waterthrush has thicker line above eye and lacks flecks on throat; prefers rocky streams.) Common to abundant winter resident, arriving in Aug. and departing by the middle of May; occurs countrywide at low and middle elevations. Winters in wet areas, especially mangroves, banks of rivers and streams, swamps, and moist agricultural areas, but during migration it can occur almost anywhere. Solitary and territorial. Walks on the ground in search of insects; flies up into vegetation when startled. Constantly bobs tail up and down. Loud harsh metallic *tsink* call is a clear indication of its presence.

## Wilson's Warbler (*Wilsonia pusilla*)

4.25 in (11 cm). Small. Yellow below, olive-green above; male has black cap, female and young have olive cap. Abundant winter resident, arriving in mid-September and departing by mid-May; occurs countrywide at middle and upper elevations, but during migration it can show up just about anywhere, including lowlands. Occupies all levels of canopy; found in forest, woodland, scrubby páramo, coffee plantations, and abandoned fields. Territorial when wintering, though it is a common member of high-elevation, mixed-species flocks. Seems constantly in motion, flicking its wings and flipping its tail as it sallies to catch bugs in the air or off vegetation. Very vocal; call is a sharp dry *jimp*, uttered every few seconds.

adult
male

adult
female

## Slate-throated Redstart (*Myioborus miniatus*)

4.75 in (12 cm). Slate gray overall; dull rufous crown patch; yellow belly; underside of tail shows white edges and tips. Juvenile has less white on tail; sooty black overall with buff belly (cf. Collared Redstart). Common; occurs on both slopes of mountain ranges at middle and upper elevations. Favors wet mountain forests but also occurs in open woodland. Most active at middle levels but sometimes ascends to canopy or descends to ground. Frequently in pairs and also regularly accompanies mixed-species flocks. Foraging behavior is distinctive: with wings partially spread, birds hop and flit, move body from side to side, and flash tail open and shut (or fan it). Flashes of white in tail cause insect prey to flush.

## Collared Redstart (*Myioborus torquatus*)

5 in (12.5 cm). Slate-colored upperparts; rufous crown patch; yellow face and throat; dark breast band; tail has white edges. Young very dark with only traces of yellow on belly. Common; occurs from Tilarán Mountain Range south to Panama, at high elevations (generally much higher than Slate-throated Redstart). Endemic to Costa Rica and Panama. Favors highland forests. Foraging behavior similar to that of Slate-throated Redstart. Pairs sing a duet (a rare trait in warblers).

## Buff-rumped Warbler (*Phaeothlypis fulvicauda*)

5 in (13 cm). Buff-yellow patch at the base of tail and rump; pale line above eye; olive-brown above. Common; occurs in Caribbean and S. Pacific at low and middle elevations. Mostly terrestrial. Favors swamps and forested rivers and streams. Usually in pairs. Hops along shady banks or trail edges in search of small insects. Can be quite tame and will sometimes allow a close approach before flying ahead a short distance. Wags spread tail to show off rump patch. Places well-camouflaged dome nest on a sloping rock or embankment. Song is a series of sharp whistles that crescendo, becoming very loud.

**GENERA INCERTAE SEDIS (Bananaquit).** The Bananaquit appears to be related to both warblers and tanagers but its taxonomy is currently undecided.

## Bananaquit (*Coereba flaveola*)

3.5 in (9 cm). Small. Short tail and legs; olive-gray above, yellow below; black head, broad white line above eye, and gray throat; bill with slight curve, thin tip. Sexes are identical; young are a very drab version of adults. Common to abundant; occurs in Caribbean and S. Pacific at low and middle elevations. Inhabits mature forests, woodlands, plantations, and gardens. Very active. At middle and high levels of forest (lower in open habitats), scrambles over vegetation to get at fruit and flowers. Probes smaller flowers (and pierces the base of larger ones) to obtain nectar; also visits hummingbird feeders. Nest is a tight ball of fibers woven into supporting vegetation; has a side entrance. Song is a series of scratchy, buzzy sputters and trills.

**Family THRAUPIDAE (Tanagers).** These are some of the showiest, most brightly colored birds of the New World. The majority occur in tropical climates. There are 42 species in Costa Rica, only 3 of which are North American migrants. Arboreal; many are core members of mixed-species flocks in forests. Some eat primarily fruit and nectar, others prefer insects. Songs are generally weak and unmusical.

### Common Bush-Tanager (*Chlorospingus ophthalmicus*)

5.25 in (13.5 cm). Largish brown head, white spot behind eye, gray throat; short bill; olive-green above, yellowish below. Abundant; occurs on Caribbean and Pacific slopes at middle and high elevations. (At higher elevations it is replaced by the equally abundant Sooty-capped Bush-Tanager, distinguished by jagged white line on blackish head.) Frequents upper levels of mature forest and nearby open areas with tall trees. Occurs in vocal, energetic groups of a half dozen or more; groups commonly lead mixed-species flocks. Eats fruit and nectar; also rummages through epiphytes and leaves in search of insects. Calls consist of excited twitters and quick, sharp, metallic notes.

### Passerini's Tanager (*Ramphocelus passerinii*)
### Cherrie's Tanager (*Ramphocelus costaricensis*)

6.25 in (16 cm). Once considered a single species, the Scarlet-rumped Tanager. Passerini's occurs on the Caribbean slope; Cherrie's is found on the S. Pacific. Males of both species are black with a brilliant scarlet rump patch. Female Passerini's: olivaceous overall; grayish head; buff throat; ochraceous-olive upper breast and rump. Female Cherrie's: similar but upper breast and rump are a dull orange. Both species are common and widespread at low and middle elevations. Inhabit second growth, scrubby fields, yards, and gardens. Usually move about in loose flocks of 6 to 10, although pairs are monogamous and territorial. Eat mostly fruits, some arthropods. Song is composed of several sweet clear notes that are repeated slowly. Calls are short and harsh; flocks maintain a constant chatter. Battling males display fluffed-out rump patches to each other.

female
Passerini's
Tanager

male

female
Cherrie's
Tanager

### Blue-gray Tanager (*Thraupis episcopus*)

6 in (15 cm). Blue wings and tail; pale blue-gray head and body. Sexes are identical. Common; occurs throughout the country, absent only at the highest elevations. Favors woodlands, farms, gardens, and parks; occasionally found in canopy or at edge of mature forests. Tolerates close association with people. Almost always seen in pairs, though noisy congregations of several-dozen birds occasionally descend on fruiting trees. Eats fruit primarily; also some insects. Cuplike, mossy nest placed in trees at medium height (less often on structures). Rambling song consists of squeaky notes and chirps. Common name in Costa Rica is *viuda* (widow), in reference to the blue-gray shawls traditionally worn by widows.

### Palm Tanager (*Thraupis palmarum*)

6.25 in (16 cm). Grayish olive with contrasting blackish wings and tail; pale wing band visible in flight. Sexes are similar. Common; occurs throughout country except at high elevations. Inhabits open areas such as woodlands, farms, gardens, and parks. Eats fruit and small arthropods. As name implies, this species is fond of palm trees; a single pair is often found in the vicinity of palm groves; prefers to nest in palm crowns amid clustered bases of fronds; incorporates palm fibers and grasses into its compact, cuplike nest. Song consists of rapid metallic notes that often rise in pitch and intensity and then fall off slightly.

### Summer Tanager (*Piranga rubra*)

6.5 in (16.5 cm). Bone-colored bill; adult male is entirely red (young male in molt is a patchy red and yellow); female and young are yellow. (Hepatic Tanager, uncommon, is darker red and has a blackish bill.) A common migrant and winter resident, arriving in Sept. and departing in April; occurs countrywide though scarce in dry N.W. Pacific and at high elevations. Inhabits forests, mangroves, open woodlands, plantations, and gardens; for many Costa Ricans, this is a familiar backyard bird. Alone or in mixed-species flocks. Moves deliberately through middle and upper levels of trees in search of fruit and insects; occasionally tears open a wasp nest to get at larvae. When agitated shows a slight crest and cocks tail. Call is a hard dry *pituk* (or *pitituktuk*).

female

male

### Golden-hooded Tanager (*Tangara larvata*)

5 in (13 cm). Small. Mostly black; adorned with iridescent gold and turquoise; belly and undertail white. Sexes similar. Juveniles are much duller (and, when very young, grayish overall). Common; occurs on Caribbean slope and in S. Pacific at low and middle elevations. Favors open areas with trees; also found in canopy of dense forest. Generally in pairs, less often in small family groups. Eats a variety of fruit; also searches for insects. Song is a repeated series of scratchy, metallic notes. Local names are *mariposa* (butterfly) and *siete colores* (seven colors).

### Bay-headed Tanager (*Tangara gyrola*)

5.25 in (13.5 cm). Small. Green above, turquoise blue below; rufous head. Sexes are alike. Young are dingier versions of adult. (Rufous-winged Tanager, a much rarer bird, overlaps partially on Caribbean slope; male Rufous-winged is similar to Bay-headed but note rufous wings, green underparts.) Common; occurs on Caribbean and in S. Pacific at middle elevations (in lowlands of the S. Pacific, occurs locally). Frequents canopy and middle levels of wet forest; also inhabits trees and shrubs in open countryside adjacent to forest. Pairs (or small groups) forage alone (or in mixed-species flocks) in search of fruit and insects. Calls are hissing *chirp* notes. Compared with other *Tangara* tanagers, the Bay-headed is uncharacteristically musical; it sometimes sings a brief lilting song (chiefly at dawn).

### Silver-throated Tanager (*Tangara icterocephala*)

5 in (13 cm). Golden-yellow head and underparts; dark stripes on back; wings and tail with greenish edges; silver-gray throat. Females and young are similar to males but duller. Common to abundant; occurs on both slopes at middle elevations; may descend to lowlands (most commonly on Caribbean slope) during prolonged rains. Favors wet mountains; inhabits middle and upper levels of forests, forest borders, and nearby open areas with scattered trees. In pairs or small groups. Active and inquisitive; birds hop on branches and foliage, peering about for insects and fruit. Frequently mixes with other species at fruiting trees; also joins foraging flocks. Call is a buzzy, sizzling *bzzeet*.

### Green Honeycreeper (*Chlorophanes spiza*)

5 in (13 cm). Male a striking blue-green overall; black hood; red eyes; curved bill is yellow and black. Female olive-green above, lime green below; lacks black hood; bill is duller than male's. Common; occurs on Caribbean and S. Pacific in wet lowlands and middle elevations (scarce and local in Caribbean lowlands). Inhabits canopy of forests, forest edges, and tall trees and shrubs in nearby open areas. Single birds or pairs frequently join mixed-species flocks. Eats fruit, flower nectar, and the occasional insect. One of the more common of its varied calls is a sharp *chip*, repeated tirelessly from concealed canopy-perch.

male

female

### Red-legged Honeycreeper (*Cyanerpes cyaneus*)

4.5 in (11.5 cm). In all plumages: red legs, slender decurved bill, and yellow underwing (visible in flight). Breeding male is violet-blue overall; powder-blue crown; black back patch, wings, and tail. Nonbreeding male: between June and Dec., male molts into green plumage but retains black wings and tail (one of just a handful of species in Costa Rica in which male molts into nonbreeding plumage). Female is olive-green overall, paler and faintly streaked below; shows a pale line above eyes. Common; occurs countrywide at low and middle elevations. Favors open country; regularly visits forests only in dry N.W. Pacific or where there are mangroves. In small groups, sometimes in pairs. Within canopy and below, explores trees and tall shrubs to ob-tain insects, fruit, and flower nectar. Flocks file out of treetops on rapidly beating wings. Call is a thin, nasal hiss.

breeding male

female

underwing

# Mixed-species Foraging Flocks

Green Honeycreeper (*Chlorophanes spiza*).

The Green Honeycreeper is one of many tropical-forest birds that travels in mixed-species foraging flocks. It joins tanagers, warblers, and other small birds in search of fruiting trees. Once such a flock locates a fruit-bearing tree, the birds alight and suspend their travels, for the moment at least. Insectivorous flocks, on the other hand, are ceaselessly on the move in search of prey. Antbirds, for example, join with other species to attend army-ant swarms, where they eat the insects that the ants flush out. As the ants move through the forest, so too the birds.

Mixed-species flocks are often noisy and conspicuous. Though they can attract the attention of predators, their large numbers mean more sets of eyes are available for spotting potential danger—and a larger flock is able to locate food more quickly. These flocks also attract the attention of birders, as they provide an opportunity to see many species at a single location.

**Family EMBERIZIDAE (New World Sparrows).** The majority of the 37 species in Costa Rica are residents. Most have stout, conical bills adapted for eating seeds. They forage on the ground or in low vegetation. Sexes are similar in the ground-sparrows and brush-finches; sexes differ in the seedeaters and grassquits. Species with melodious songs are popular cage birds.

## Blue-black Grassquit (*Volatinia jacarina*)

4 in (10 cm). Adult male is a glossy blue-black; in flight, some white is visible on underside of wing. Female and young are brownish with dark streaking. Bill conical. (Caribbean race of male Variable Seedeater is less glossy, shows white spot on folded wing, and has a rounded bill.) Common; occurs countrywide at low and middle elevations. Favors weedy fields, overgrown pastures, roadsides with tall grassy margins, and plantations (of sugar cane and rice). Sometimes forms large flocks, mixing with other species, but also found alone. From a slightly elevated perch, male tirelessly performs, making vertical leaps while emitting buzzy calls. Called Johnny-jump-up within English-speaking regions of the Caribbean.

female

male

## Variable Seedeater (*Sporophila americana*)

4 in (10.5 cm). Stubby, curved bill is characteristic of the genus. Males of both races show a white spot on folded wing. Female is brown to olive. Common to abundant; occurs on Caribbean and Pacific slopes at low elevations (though rare in dry N.W. Pacific). Prefers low second growth, overgrown grassy or weedy pastures, scrubby forest edges, and gardens. Usually alone or in small groups but sometimes joins other species in areas with seeding grass. Clings to grass stems when eating seeds. From a high perch, male sings a rushing, twittery song that sometimes includes phrases from other species.

male
Caribbean
race

female

male
Pacific race

### Yellow-faced Grassquit (*Tiaris olivaceus*)

4 in (10 cm). Mostly olive; small, conical bill is grayish. Male has yellow line above eyes and yellow on throat; black face and chest. Female and young are olive or gray with indistinct facial markings. Common; occurs on both slopes at middle elevations (and above); also found occasionally at low elevations (though absent from the dry N.W. Pacific). Inhabits pastures, lawns, weedy fields, and grassy roadsides. Usually in pairs or small groups; joins other species where seeds are abundant. Even though song is a weak insectlike trill, this is a favorite cage bird. In Spanish known as *gallito* (little rooster), because males quiver wings while delivering song.

male

female

### Slaty Flowerpiercer (*Diglossa plumbea*)

4 in (10 cm). Little. Upturned, hook-tipped bill. Male entirely slate gray. Female and young are brownish and faintly streaked. Abundant; occurs on all major mountain ranges from high elevations to above tree line. Endemic to Costa Rica and Panama. Inhabits páramo, forest, forest edges, scrubby clearings, and gardens. Forages at all levels, alone or in pairs. Restless; constantly flicks wings and nervously flits through foliage; makes brief visits to flowers or conducts short sallies to catch flying insects. Flowers are punctured with the pointed lower mandible and nectar is extracted with the aid of a brush-tipped tongue. The hooked upper mandible is used only to grasp flowers.

male

female

## Yellow-thighed Finch (*Pselliophorus tibialis*)

7.25 in (18.5 cm). Puffy yellow thighs; long tail; black cap, wings, and tail; gray-olive underparts. Sexes are similar. Young lack yellow thighs. Common, occurs in Central, Tilarán, and Talamanca mountain ranges at high elevations. Endemic to Costa Rica and Panama. Prefers thickets, tangled clearings, and lower levels of dense forest. Chattering pairs or small family groups often accompany mixed-species flocks led by bush-tanagers. Varied diet includes fruits, insects, and nectar. In deep shade, often only the yellow thighs are visible, creating the illusion of a pair of hopping, yellow spots. Pairs communicate with harsh and twittering calls; male's song is short and melodious.

## Orange-billed Sparrow (*Arremon aurantiirostris*)

6 in (15.5 cm). Bright-orange bill; thick white line above eye; black breast band; olive upperparts. Young have a dark bill and are duller overall. Common; occurs on Caribbean in wet lowland forests; occurs in S. Pacific at low and middle elevations. Frequents forest interior and nearby second growth. Activity confined to forest floor and low, dense vegetation. Usually in pairs. Sometimes skulks but also seen at trail edges hopping and scratching in search of insects and fallen seeds. Bulky dome-shaped nest, composed of leaves and fibers, is placed on ground; has side entrance. Call is a high-pitched, flinty *chip*. Both sexes perform song, a series of high, tinkling, metallic notes.

## Black-striped Sparrow (*Arremonops conirostris*)

6.5 in (16.5 cm). Chunky. Olive-green on upperparts and section of underparts; gray head with bold black stripes; white throat and belly; trace of yellow on bend of wing. Sexes similar. Young darker and faintly streaked. Abundant; occurs on Caribbean and on S. Pacific at low and middle elevations. Inhabits second growth, forest edges, overgrown fields, plantations, and shady yards and gardens. Pairs or family groups forage for insects, seeds, and berries; generally on the ground or in low tangled vegetation (rarely ventures higher except to sing). More often heard than seen, though it is somewhat curious and may pop out briefly to glimpse observers. On reuniting, both pair members utter jumbled, excited, slurred whistles. Sings from a concealed perch; song is an accelerating series of clear whistles (often the first bird to sing at daybreak).

### Stripe-headed Sparrow (*Aimophila ruficauda*)

7 in (18 cm). Bold black stripes on head; white throat; upperparts tawny with brown streaks; mottled gray below; long, cinnamon-colored tail. Common and conspicuous; occurs in dry N.W. Pacific at low and middle elevations. Inhabits brushy or grassy open-country habitat. Noisy cohesive groups of up to a dozen birds hop on ground and through vegetation in order to gather seeds and insects. Has weak flight; seldom airborne for more than a moment. When alarmed, birds rush collectively toward the cover of a low tree or bush, all the while emitting sputtering, squeaky calls. Occasionally appears on open perches at eye level or above.

### Rufous-collared Sparrow (*Zonatrichia capensis*)

5.25 in (13.5 cm). Gray-and-black head with short crest; rufous collar; white throat, grayish underparts. Young lack rufous collar and have faint streaking on breast. Common to abundant; occurs countrywide at middle and high elevations. Found in a variety of open habitats, including páramo, pastures, brushy fields, plantations, and gardens. Usually in pairs. Hops on ground foraging for seeds and insects. Male sings from an elevated perch; song varies greatly from location to location. Local name is *comemaiz* (corn eater).

young

adult

### Volcano Junco (*Junco vulcani*)

6.25 in (16 cm). Yellow eyes; pale-pink bill; brownish streaking on back, wings, and tail; gray below. Fairly common; occurs on highest peaks, from Irazú Volcano south to the Talamanca Mountain Range; generally found above tree line though it sometimes descends to lower elevations in deforested areas. Endemic to Costa Rica and Panama. Frequents clearings in páramo, scrubby pastures, and roadsides. Alone or in pairs. Fairly tame and lethargic; hops over open ground in search of seeds, berries, and insects; flies lazily up into bushes when flushed.

**Family CARDINALIDAE (Saltators, Grosbeaks, and Buntings).** Closely related to the Emberizidae (New World Sparrows). There are 14 species in Costa Rica, 8 resident. Most have large, stout bills; many sing melodious songs. In tropical species, sexes tend to be alike; migrants show sexual dimorphism.

## Buff-throated Saltator (*Saltator maximus*)

8 in (20 cm). Bright olive-green above; grayish head and underparts; buff throat patch bordered by broad black bib. Of the four saltators that occur in Costa Rica—all of them resident—this is by far the most common and widespread species. Occurs throughout the country at low and middle elevations (except in dry N.W. Pacific, where scarce and local). Pairs inhabit tall second growth, forest edges, brushy fields, plantations, and gardens with trees. Forages at middle level of canopy for insects, fruits, and nectar; when foraging, mixes with other open-country birds. Seems coy, craning about to peer at observer through gaps in foliage before flitting away and making high-pitched, tittering calls. Quite vocal; song is composed of a prolonged lilting series of sweet whistles.

**Family ICTERIDAE (Blackbirds and Orioles).** A large, diverse New World family. Most species have a pointed, conical bill. Many are predominantly black. Males are often bigger than females. There are 23 species in Costa Rica; the majority are breeding residents, 5 are migrants from North America.

## Red-winged Blackbird (*Agelaius phoeniceus*)

Male 8.75 in (22 cm); female 7 in (18 cm). Red-and-yellow epaulet on wing of adult male is diagnostic. Female and young blackish brown with substantial streaking above and below. Abundant but local; occurs at low elevations in northern half of country. (On both slopes, slowly moving south in the wake of deforestation.) Inhabits marshes, brushy wet pastures, and agricultural lands. Forages on the ground for seeds and insects. Highly gregarious; when feeding and roosting, sometimes forms huge flocks; nests in colonies (in inundated reeds or brush). Defends its nest fearlessly and tenaciously. Male perches on stem, fence, or utility wire to sing (and simultaneously display his epaulets). Song is composed of three or four harsh gurgling whistles.

female

male

### Great-tailed Grackle (*Quiscalus mexicanus*)

Male 17 in (43 cm); female 13 in (33 cm). Male has iridescent purplish-black plumage; long, keeled tail; yellow eyes. Female is dark brown above, buff brown below; pale line above eye. Originally restricted to the coasts, this species has followed humans inland, even up to high elevations, and is now common countrywide in both agricultural and urban settings. Alone or in loose flocks. Bold and aggressive. Eats a wide variety of plants and animals, including nestlings and eggs of other birds; also scavenges on beaches and in towns. Massive flocks roost in city parks, creating a nuisance with droppings and deafening cries. In aggressive displays, male stretches out his body and points bill skyward. Song consists of shrill rising whistles or soft, surprisingly melodious, phrases.

female

male

### Bronzed Cowbird (*Molothrus aeneus*)

8 in (20 cm). Chunky build; fairly short bill, red eyes. Male is bronzy black with a greenish gloss; female is dull brown; young birds have dark eyes. Common; occurs throughout the country at low and middle elevations. Favors humid, deforested regions (scarce and local in dry zones). Found in open country, yards, gardens, and agricultural areas, especially where there is livestock. In large or small cohesive flocks most of the year. Forages on ground for grains, seeds, and insects. Lays eggs in nests of other bird species. During display ritual, male (sometimes several) puffs out neck feathers and hovers a few inches off the ground in front of the selected female. Male song repertoire includes squeaky, wheezy whistles and soft rattles.

young

male

## Black-cowled Oriole (*Icterus prosthemelas*)

7.5 in (19 cm). Adults are black on chest, upperparts, and tail; yellow belly, rump, and wing patch. Young are olive-yellow above. (Rare Yellow-tailed Oriole, of Caribbean lowlands, shows yellow in tail.) Common; occurs on Caribbean slope at low and middle elevations. Inhabits open country with scattered trees, forest edges, banana plantations, gardens, and yards; often found close to water. Usually in busy pairs. Hops and rummages acrobatically through foliage of palms, bananas, and other small trees in search of fruit, nectar, and insects. Emits a constant, scratchy, inquisitive call. It sings a plain song and is therefore not a popular cage bird, unlike more melodious species of oriole.

adult

young

## Streak-backed Oriole (*Icterus pustulatus*)

7.5 in (19 cm). Orange to orange-yellow; dark streaks on back; black throat and tail; wing feathers are black with white margins and tips. Young are duller and above show an olive wash. Common; occurs in dry N.W. Pacific at low and middle elevations. Inhabits forests, scrubby pastures, and yards. Pairs or family groups forage in upper layers of trees, shrubs, and vines. Eats insects, fruit, and flower nectar. Pendent nest, built of densely woven black fibers, is often placed in a Bull-horn Acacia, a tree that is invariably filled with stinging ants. Calls can be harsh or musical; also makes a dry chatter. Whistled song is sweet and subdued.

## Baltimore Oriole (*Icterus galbula*)

7 in (18 cm). Adult male is distinctive: black upperparts and bright-orange underparts, white wing bars. Female and young are clad in dull shades of orange, brown, and gray. A common North American migrant (and winter resident), arriving in Sept. and departing in early May. Occurs countrywide (though scarcer in highlands). Frequents forest, forest edge, open areas with trees, agricultural lands, gardens, and yards. Usually in small groups that mix freely with other species. Eats mostly fruit and nectar, but also searches foliage and tangled vegetation for insects. This showy bird can be amazingly difficult to spot when in an orange-flowered *Erythrina* tree, a favorite source of nectar. Frequent call is a dry sputtering rattle, though it occasionally emits one or two sweet whistles.

male

female

## Scarlet-rumped Cacique (*Cacicus uropygialis*)

9 in (23 cm). Black with a red rump patch that is often concealed; pale-blue eyes; whitish, pointed bill. Occurs on Caribbean slope and S. Pacific at low elevations; locally, ventures up into foothills. Common in undisturbed areas; scarce or absent in deforested regions. Found in forest and nearby open areas with tall trees. Small flocks swoop noisily through canopy, landing to search epiphytes and other vegetation for large insects or to eat fruit and nectar. Inserts closed bill into object of interest and then opens bill to pry it apart. Forms mixed-species flocks with other large birds. Calls constantly; emits rich throaty whistles, some strident, others questioning. Pairs nest alone; often suspends pouchlike nest near predatory wasps that feed on the botflies that parasitize nestlings.

## Montezuma Oropendola (*Psarocolius montezuma*)

Male 20 in (50 cm); female 15 in (38 cm). Chestnut body; black head and neck; bill has black base, orange tip; blue skin on face; yellow tail with dark central feathers. Very common; occurs at low and middle elevations on Caribbean slope; occurs at middle elevations on northern section of Pacific slope. Found in forest, broken woodland, and agricultural regions with tall trees. Mostly in canopy, but also lower. Bounces heavily over branches and rummages through bromeliads and palms in search of insects, small animals, fruit, and nectar. Large flocks skim treetops commuting to and from feeding areas. Breeds in colonies; usually selects a tall isolated tree, the outer branches of which it festoons with pendulous, woven-fiber nests. Makes querulous calls. When displaying, male tips forward on perch and flutters wings and tail, uttering bizarre rasping vocalization.

# Safety in Numbers

Montezuma Oropendola (*Psarocolius montezuma*) and nests.

The Montezuma Oropendola is a colonial breeder. Groups build nests in a single tree, displaying an evident preference for nesting very close together; as many as 21 large, hanging nests are sometimes placed on one branch—even when empty branches are available. This is risky business, however, as heavily laden branches can snap off. Clustering of nests perhaps enables females to more easily gang up against predators such as snakes and hawks, and also against the Giant Cowbird, a nest parasite.

The Giant Cowbird (*Molothrus oryzivorus*) sneaks in, lays a single egg in an oropendola nest, and then flees; if a cowbird egg goes unnoticed, the oropendola mother unwittingly takes on the job of incubating it, a waste of precious resources. But where nests are clustered, the odds are increased that at least one female will be present to drive away the intruder.

Nest clustering offers yet another advantage to females, whose preference is to mate with an alpha male. When subordinate males approach with designs of finding a mate, the females can collectively drive off the unwanted suitors.

155

**Family FRINGILLIDAE (Euphonias and Finches).** In Costa Rica there are 10 euphonia species and 2 finch species, all resident. Euphonias, until recently placed in the tanager family, have short bills, legs, and tails. Males are brightly colored; females are usually clad in drab yellows and olive greens. Distinguishing between species is often a challenge. Euphonias are able to eat a favorite food—the toxic mistletoe berry—because their rudimentary gut passes it quickly.

## Yellow-crowned Euphonia (*Euphonia luteicapilla*)

3.75 in (9.5 cm). Male is glossy blue-black above, bright yellow below except for dark throat; large yellow crown patch. Female is olive green above, pale yellow below. The most widespread euphonia in CR; occurs on Caribbean and Pacific slopes at low and middle elevations. Common in humid regions, rare and local in the dry N.W. Pacific, where it is replaced by the Scrub Euphonia. Prefers open country with scattered tall trees, though it sometimes visits forests. Pairs or small groups frequent canopy in search of berries and other fruits; especially fond of mistletoe berries and figs. Often perches in bare tree crown, yet can be inconspicuous when feeding. Call consists of clear strident whistles: two high notes then a pause succeeded by three low notes.

## Yellow-throated Euphonia (*Euphonia hirundinacea*)

4.25 in (11 cm). Both sexes have a white belly. Male is glossy blue-black above, bright yellow below; small yellow crown patch. Female is olive green above and pale yellow below. Common; occurs in northern half of country, principally at low and middle elevations on the Caribbean slope and at middle elevations on the Pacific slope. In S. Pacific, replaced by Thick-billed Euphonia. Found in disturbed forest, forest edges, open woodlands, plantations, and yards. Usually in pairs, less often in small groups. Forages in trees for fruits and insects. A skilled mimic of the simple calls of other birds.

### Golden-browed Chlorophonia (*Chlorophonia callophrys*)

5 in (13 cm). Large, plump. Bright-green plumage; baby-blue crown; yellow belly. Male is brighter than female and has long yellow eyebrow and narrow black chest band. Common; occurs in all the major mountain ranges at middle and high elevations. Moves to lower elevations during the rainiest months of the year (a common trait of highland birds that eat fruit). Endemic to Costa Rica and Panama. Favors canopy of wet mountain forests and nearby open areas with tall trees. In pairs or small groups. Mainly eats small fruits of trees and epiphytes. Though it vocalizes constantly, often difficult to detect because of green plumage and rather sedate habits. Calls include soft mournful whistles and nasal clucks.

male

female

# reptiles

**There are more than** 8,200 species of reptile in the world. Costa Rica is home to 222 species divided into three orders that are only distantly related: crocodilians, turtles, and squamates (lizards and snakes). Although many reptiles are nocturnal and elusive creatures, an astute observer on even a short walk through the rainforest or by a stream is almost certain to see a beautiful or interesting representative of their kind.

All crocodilians are assigned to the archosaurs, an ancient group of reptiles closely related to birds. A large population of American Crocodiles, some of them over 16 feet long, inhabits the Tarcoles River. The best place to see these is from the bridge near Carara National Park, although visitors to Tortuguero National Park, the Osa Peninsula, Manuel Antonio, and any number of other places in Costa Rica, are also likely to see crocodilians.

Fourteen species of turtle occur in Costa Rica. Six of these are large majestic species of marine turtle that nest on the country's beaches. Witnessing their arrival to lay eggs is to experience one of the world's great natural history dramas. A number of beaches are ideal places to see nesting turtles, including Playa Grande and Gandoca-Manzanillo (Leatherback Turtles), Santa Rosa National Park (Olive Ridleys), and Tortuguero National Park (Atlantic Green Turtles and other species). Marine turtles only nest at certain times of the year, so it is important to do a little research before making plans to see them.

By far the largest group of reptiles is the order Squamata, which includes both the lizard and snake suborders. Costa Rican lizards—there are 69 species—range in size from tiny geckos to the enormous Green Iguanas that lounge in trees near rivers. Anoles are present just about wherever there is vegetation. Male anoles, like females, are generally cryptic green or brown, but the males occasionally break their camouflage by flashing a brightly colored dewlap to assert their territorial rights or to court a female. Several species of basilisk lizard inhabit the banks of lowland rivers. These speedy reptiles are nicknamed Jesus Christ lizards because of their ability to run across the surface of water in short bursts.

No less than 137 species of snake inhabit Costa Rica. The Boa Constrictor and other large, powerful snakes occur here, but so too do relatively small snakes that eat invertebrates, swallowed alive. So-called goo-eaters, snakes that feed on snails, slugs, and other soft-bodied prey, are common. Roughly half of the country's snakes possess enlarged teeth or fangs, many of them with glands that secrete toxins. A mere 20 species, however, are considered dangerously venomous, among them the coral snakes, Fer-de-Lance, Eyelash Pitviper, and, of near mythical status, the Bushmaster. A good way to spot snakes and lizards is to look for a sunny patch in the forest, where they commonly bask, either early in the day or after a rainy spell.

Cloud Forest Parrot Snake (*Leptophis nebulosus*). This diurnal snake hunts for sleeping tree frogs hidden in low vegetation in its cloud forest habitat. When threatened, it flashes its colorful mouth lining.

## Order Crocodylia (Caimans and Crocodiles)

Large, distinctive creatures with serrated keels on back and tail. Rely on powerful limbs and laterally compressed tail when swimming. Two species occur in Costa Rica, each in its own family: the Spectacled Caiman in Alligatoridae and the American Crocodile in Crocodylidae. Eggs are incubated by heat generated from nest's decomposing vegetation. Sex of offspring determined by incubation temperature. Young fall prey to many predators, but adults have no natural enemies apart from humans. Large crocodiles sometimes attack humans.

## Spectacled Caiman (*Caiman crocodilus*)

To 8.9 ft (2.7 m), but usually considerably smaller. Bony ridge in front of eyes is diagnostic. Tan, gray, or brown. Young more vividly marked. Can be common locally; occurs on Caribbean and Pacific coasts in lowland rivers, swamps, and ponds. During the day, basks on river banks, sandbars, and exposed logs. At night trolls water for invertebrates and small fish and amphibians. Generally shy and elusive. Most frequently observed during dry season, when it concentrates in shrunken bodies of water; at night, a powerful flashlight trained at a body of water often reveals eyeshine from multiple caimans. Breeds during the rainy season; deposits 15 to 40 elliptical, hard-shelled eggs in mound-shaped nest (a low heap of vegetation and leaf litter) located above high-water mark on riverbank. Parents guard nest.

## American Crocodile (*Crocodylus acutus*)

To 24 ft (7.3 m), but females and most males under 13 ft (4 m). Snout elongate and slender. Grayish green, dark olive green, or brownish gray; with dark bands across the back and tail. Young more boldly marked than adults. Occurs on both coasts in lowland rivers, swamps, lagoons, and estuaries; can be common locally. Usually seen basking on shore or on sandbanks, lying motionless for hours. Easy to spot from the bridge that spans the Tarcoles River. When body temperature climbs too high, enters water or opens mouth to cool down. Excellent swimmer; large individuals can remain submerged for more than one hour. Feeds on variety of aquatic animals, including crabs, amphibians, and fish; infrequently, eats water fowl or small mammals. Adults sometimes excavate river-bank burrows with underwater entrance. Toward the end of the dry season, female deposits 20 to 60 elliptical eggs in a shallow nest excavated near water.

## Order Testudines (Turtles)

There are 14 species in Costa Rica, including both marine and freshwater varieties. Turtles lack teeth, biting instead with sharp, keratin-covered jaw sheaths. Some turtles are vegetarian, some carnivorous, and others omnivorous. All turtles lay eggs; in many species, the gender of the offspring is determined by the incubation temperature of the eggs. Turtles may live for several decades. Sea turtles have paddle-shaped limbs that they use to swim long distances. Adult sea turtles only come ashore to nest, returning to deposit their eggs at the very same beach where they themselves hatched. Freshwater turtles have well-developed toes that can be extensively webbed in some aquatic species. They occur in a variety of habitats, sometimes a considerable distance from water.

All turtle measurements indicate the length of the shell (tail not included).

**Family CHELONIIDAE (Hard-shelled Sea Turtles)**. Five species of hard-shelled sea turtle are known to lay their eggs on Costa Rican beaches. All have a hard, bony shell.

## Olive Ridley (*Lepidochelys olivacea*)

28 in (0.7 m). The smallest sea turtle in Costa Rica. Shell heart-shaped with slight keel; olive-colored above and lighter below. Male has long tail that extends far beyond hind margin of shell; female has shorter tail. Occurs on Pacific beaches, often in very large numbers. In Guanacaste Province, thousands of Olive Ridleys sometimes nest on the beaches of Ostional and Nancite (such mass arrivals are called *arribadas* in Spanish). At Ostional Beach, somewhere between 150,000 and 200,000 turtles may arrive over the course of just a few days; the beach becomes so crowded that new arrivals end up destroying previously laid eggs in the process of depositing their own. Nesting takes place at irregular intervals throughout the rainy season, with peak activity in Sept. or Oct. Female usually deposits over 100 eggs per visit; incubation generally takes between 50 and 70 days. On land and in the ocean, a variety of predators prey on hatchlings, and few survive. Costa Rican law allows limited, sustainable harvesting of eggs during the first hours of an *arribada*. *Illustration not to scale.*

female

## Atlantic Green Turtle (*Chelonia mydas*)

60 in (1.5 m); weighs over 605 lbs (275 kg). Very large. Broad, smooth, heart-shaped shell; mottled olive to dark brown. Male has a single, curved claw on each front flipper that is used to cling to female's shell during mating. (Similar Pacific Green Turtle, *Chelonia agassizii*, occurs only on Pacific beaches.) Occurs on Caribbean beaches; there is a large population in Tortuguero National Park. Mates in shallow waters off nesting beaches, usually close to shore. Gravid females come ashore roughly between June and Sept., every two to four years; during the breeding season, they typically produce multiple clutches of about 100 eggs each. Incubation time is from 50 to 70 days. Eggs and hatchlings preyed upon by many animals and early survival rate is thus extremely low. Young initially eat small invertebrates but gradually shift to vegetarian diet.

**Family DERMOCHELYIDAE (Leatherback Turtle).** The lone species in its family, the Leatherback Turtle is the largest reptile in the world, and the only marine turtle with a soft, leathery shell.

## Leatherback Turtle (*Dermochelys coriacea*)

72 in (1.8 m); weighs over 1,435 lbs (650 kg). Enormous. In place of hard, bony shell has leathery carapace supported by seven prominent ridges. Long, paddlelike limbs lack claws. Dark gray with whitish blotches on shell, head, and limbs. Male has long tail that extends beyond length of hind limbs. Nests on Caribbean beaches at Gandoca-Manzanillo and other locations; nests on Pacific beaches at Playa Grande, the Nicoya Peninsula, and the Osa Peninsula. Travels great distances across open ocean, migrating into cold waters to hunt jellyfish. Can dive over 3,280 ft (1,000 m) in pursuit of prey. Long, spinelike projections on upper jaws and in esophagus help clutch and swallow slippery prey. Many die due to ingestion of discarded plastic bags that are mistaken for jellyfish.

female

# Sex Determination in Turtles

Atlantic Green Turtle (*Chelonia mydas*) in Tortuguero National Park.

One of the most evocative sights in Costa Rica is the arrival of nesting turtles to its shores. The female Green Turtle normally performs the entire egg-laying ritual at night, returning to the water long before sunrise, though sometimes, as in this photo, she returns shortly after dawn. A female lays clutches of about 100 eggs, and generally produces more than one clutch during the breeding season.

In nearly all vertebrates, the sex of offspring is determined genetically, but the Green Turtle (and some other reptiles) is a fascinating exception to the rule. The sex of the young turtles is determined by the incubation temperature of the eggs. Eggs placed at the bottom of the nest or near its center, where temperatures are lower, result in male young. Eggs closer to the warm surface of the nest result in females. Each clutch generally produces some males and some females, ideally at a one-to-one ratio. In some locations, however, the effects of global warming have begun to shift the sex ratio to a disproportionate number of females. Temperature-dependent sex determination also has important conservation implications for the design of turtle-hatcheries, which, if constructed improperly, will produce offspring of only one sex.

**Family EMYDIDAE (Freshwater, Marsh, and Forest Turtles).** This is the largest family of turtles. The limbs are never paddle-shaped as in marine turtles. Amount of webbing between toes varies from species to species, but is often extensive on aquatic species like the Black River Turtle. Four species occur in Costa Rica.

## Black River Turtle (*Rhinoclemmys funerea*)

14.6 in (37 cm). Large. Smooth shell has high dome and a slight keel down the center; uniform dark brown (or black) in adults, tinged with yellow in young. Long neck and relatively small head. Extensive webbing on toes and fingers. Common; occurs on Caribbean slope to 1,970 ft (600 m), in slow-moving rivers, swamps, and other freshwater habitats. Prominently basks on logs and sandbars. Skittish; when approached, dives into the water and remains submerged for several minutes, surfacing in a different spot. Ventures onto land at night, sometimes a considerable distance from water. Actively searches for grasses, leaves, fruits, and other food (perhaps an important disperser of plant seeds along water courses).

## Red Turtle (*Rhinoclemmys pulcherrima*)

Female 7.9 in (20 cm); male smaller. Medium size. Domed shell with slight keel down the center. Lacks webbing between toes and fingers. Spectacularly colored. Occurs in dry N.W. Pacific and Nicoya Peninsula to 3,770 ft (1,150 m). Often found in gallery forests, near water. Semi-terrestrial; usually encountered on land, but readily takes to water during dry spells. Diurnal. During rainy season, female lays 1 to 3 eggs, which are deposited in a depression excavated at the base of clumps of grass or between the roots of shrubs.

**Family CHELYDRIDAE (Snapping Turtles).** A small family of aggressive, prehistoric-looking turtles. With their large head and long tail, these turtles generally appear to have outgrown their shell. There is a single species in Costa Rica.

## Snapping Turtle (*Chelydra serpentina*)

19.3 in (49 cm). Large. Long neck, massive head, and a very long tail (sometimes as long as—or longer—than the shell). Shell with three low keels. Color ranges from brown to almost black; often mottled by mud and algal growth. Widespread but seen infrequently; occurs on Caribbean and Pacific coasts to 3,610 ft (1,100 m); absent from the Nicoya Peninsula and dry regions of Guanacaste. Aquatic. Inhabits stagnant bodies of water and slow-moving rivers and streams, especially those with abundant aquatic vegetation and soft, muddy bottoms. During the day, seeks warm, shallow water, where it burrows in sediment; extends long neck to breathe at water surface. At night, walks along river bottom foraging for vertebrate prey. Female lays 20 to 40 large, hard-shelled eggs in shallow nest. On land, this turtle is irrascible and quick to bite (can be very painful), but in the water tends to be docile and shy.

**Family KINOSTERNIDAE (Mud Turtles).** A family of small, highly domed turtles. There are three species in Costa Rica. A distinguishing feature is hinged belly armor that allows them to partially or entirely close the shell.

## White-lipped Mud Turtle (*Kinosternon leucostomum*)

6.7 in (17 cm). Medium size. Has a smooth, highly domed shell; hinged belly armor allows for complete closure of shell, leaving no soft parts exposed. Limbs short and robust; feet webbed. Shell is dark brown to almost black; belly armor is yellow. Common and widespread; occurs on Caribbean and Pacific slopes to 3,940 ft (1,200 m), but absent from dry regions of N.W. Pacific and the Nicoya Peninsula. Usually seen in or near streams, marshes, or ponds, but occasionally found considerable distance from water. During the day, buries itself in loose sediment in body of water; at night walks along bottom of water in search of food. Eats plant matter, algae, seeds, and small invertebrates. Mates in shallow water after elaborate courtship behavior. Twice yearly, lays 1 to 5 oblong hard-shelled eggs in shallow depression in leaf litter.

All lizard measurements are for the combined length of the head, body, and tail.

**Family EUBLEPHARIDAE (Eyelash Geckos).** Formerly considered a subfamily of Gekkonidae (Geckos). Eyelash geckos differ from other geckos in having moveable eyelids. In many species, these eyelids bear a fringe of enlarged scales that can give the impression of eyelashes. The skin is covered by small, granular scales that create a velvety appearance. Eyelash geckos lack expanded finger and toe pads.

## Central American Banded Gecko (*Coleonyx mitratus*)

7.5 in (19 cm). The only Costa Rican gecko with moveable eyelids. Pupils vertical. Fingers and toes are narrow and lack expanded pads on the tips. Pattern of wide, yellowish-tan bands on dark-brown background; most prominent in young, obscure in large adults. Occurs on Pacific slope, predominantly in N.W. Pacific and central Pacific, to 4,590 ft (1,400 m). Secretive, shy, nocturnal. At night, actively forages for small invertebrates; during day, stays hidden in leaf litter or below debris. When stalking prey, undulates tail in a catlike manner. When alarmed, inflates throat and raises its body off the ground on outstretched legs. May vocalize when handled. Deposits clutch of 2 leathery-shelled eggs below debris or in crevices.

**Family GEKKONIDAE (Geckos).** All geckos lack moveable eyelids and have small granular scales on body, head, and limbs. Two subfamilies occur in Costa Rica. Members of subfamily Sphaerodactylinae are diurnal, have round pupils, and lack widely expanded finger and toe pads (e.g., Yellow-headed Gecko). Members of subfamily Gekkoninae are nocturnal, have vertically elliptical pupils, and have widely expanded finger and toe pads that enable them to scale vertical surfaces with ease or even run upside down on ceilings (e.g., Wandering Gecko, Turnip-tail Gecko).

## Yellow-headed Gecko (*Gonatodes albogularis*)

4.3 in (11 cm). Small to medium. Male with bright orange to yellow head, light-blue spots along the upper lip, and distinct white tail tip. Female and young with brown, cream, and black mottling; usually with a thin, pale collar. Abundant; occurs on Caribbean and Pacific slopes, to 1,640 ft (500 m). Seen on walls of buildings, coconut palms lining beaches, and trees with deeply creviced bark. Diurnal, but avoids sun during hottest time of day. Extremely territorial; males posture to each other by raising body off the ground and curling the tail over the back to emphasize the white tail tip. Reproduces year-round; female lays a single, spherical, hard-shelled egg that hatches in about 4 months. Young can reach sexual maturity in 6 months.

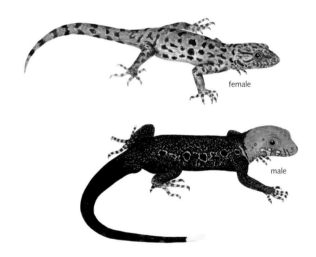

female

male

## Wandering Gecko (*Hemidactylus garnotii*)

4.7 in (12 cm). Medium size. Gray, grayish brown, or reddish brown. A dark stripe usually present behind each eye. Generally uniformly pale at night and darkly mottled during the day. Vertical pupils. Tail with a fringe of pointed tubercules along each side. Fingers and toes bear an expanded pad and claw. An introduced species. Common to abundant; occurs on Pacific slope, from central coast south, to 3,810 ft (1,160 m); rapidly expanding its range. Nocturnal. Generally encountered on walls and roofs of buildings. Sometimes produces a loud squeak when grabbed. The Wandering Gecko is an all-female parthenogenetic species: reproduction occurs without fertilization by males. *Illustration not to scale.*

## Turnip-tail Gecko (*Thecadactylus rapicauda*)

8.7 in (22 cm). Largish, robust. Large head, short limbs. Tail typically swollen at base. Expanded pads on hands and feet. Vertical pupils. Capable of considerable color change: during the day, cryptically colored in shades of brown; at night, usually pale gray or cream. Widespread along both coasts in humid habitats, to 3,450 ft (1,050 m); absent from the driest areas of the N.W. Pacific. Arboreal. Commonly found deep within pristine forests, but also occurs in people's homes. Primarily nocturnal, but sometimes hunts for large insects beginning in late afternoon or at dusk. During the day, hides below loose bark, in tree crevices, and in thatched roofs. Deposits single egg in leaf-litter accumulations. At night, produces series of loud vocalizations that are perhaps territorial. *Illustration not to scale.*

**Family IGUANIDAE (Iguanas and relatives).** The largest family of lizards in Costa Rica. An extremely diverse family, so much so that groups within the family are sometimes treated as separate families. These include the casque-headed lizards (e.g., basilisks, Helmeted Iguana); large iguanas with a serrated dorsal crest (e.g., Spiny-tailed Iguana, Green Iguana); and the anoles, the largest subgroup in the family. These are among the most frequently seen lizards in the country.

## Green Basilisk (*Basiliscus plumifrons*)

35 in (90 cm). Large. Male has sail-like crests on head, back, and tail; female with small crests, young lacks crests. Brilliant green overall with blue-and-white spots on the sides; fiery-yellow irises. Male with blue throat during mating season. Young less conspicuous. Common; occurs on Caribbean slope and in S. Pacific (in Golfo Dulce region), to 2,540 ft (775 m). Found along rivers and streams, perched on logs, rocks, and bushes. Adults often prefer shaded spots; young regularly seen in open, sunny patches, basking on rocks or logs. When approached, initially maintains frozen stance but then launches itself into water or dense vegetation. Using its hind legs, can run a considerable distance across the surface of water. Feeds mainly on insects and other arthropods, but occasionally eats small vertebrates such as lizards and even fish.

male

## Striped Basilisk (*Basiliscus vittatus*)

20 in (50 cm). Largish. Pointed, almost triangular crest on the head; low crests on the back and tail. Brown body and head; white to yellow stripes along the side. (Similar-looking Brown Basilisk, *Basiliscus basiliscus*, is larger and found only on Pacific slope.) Common; occurs on Caribbean slope to 820 ft (250 m). Often found considerable distance from water. Prefers open areas, plantations, and scrubby vegetation; commonly seen in vegetation bordering beaches. Diurnal. During day, actively hunts for invertebrate prey on ground; at night, sleeps exposed on leaves or branches that overhang water. Annually produces 1 to 4 clutches of 2 to18 eggs each. Reproductive activity starts during the second half of the dry season; most eggs hatch at the beginning of the rainy season, when insect prey is most abundant.

**Helmeted Iguana** (*Corytophanes cristatus*)

14 in (36 cm). Medium size. Prominent helmetlike crest on the head continues along the middle of the back to form a low, sawtooth ridge. Limbs long and slender; very long toes. Large, extendable throat fan. Can change its color rapidly; may be reddish brown, brown, tan, olive, or black, sometimes with irregular blotches and spots. Common; occurs on Caribbean and Pacific slopes to 5,250 ft (1,600 m); absent from dry regions of the Pacific slope. Found in humid forests. Arboreal, often motionless, and expertly camouflaged, this lizard is difficult to spot as it perches on a vertical branch or trunk. A sit-and-wait predator; in frozen stance, waits for suitable prey (large invertebrates and occasionally small lizards) to pass. Moves so infrequently that small epiphytic plants sometimes grow on its head. Female lays 5 or 6 eggs in a shallow depression on the ground, possibly using her head crest to excavate the nest. *Illustration not to scale.*

**Spiny-tailed Iguana** (*Ctenosaura similis*)

51 in (130 cm). Very large. Male with massive head, robust body, and muscular limbs; enlarged spiny scales encircle the tail in distinct bands; crest on head and back. Female smaller. Tan, olive-brown, or bluish gray, with broad, black bands. Young bright green. Abundant; occurs on Pacific slope to 2,460 ft (750 m). Found in open habitats of coastal plains, on buildings, in vacant lots, and along roadsides; often conspicuous along many beaches. Basks on sunny days (tolerates very high temperatures). An excellent climber; perches on branches high in trees. Young mostly active on ground. Extremely territorial; male performs ritualized head-bobbing display to drive off intruders or attract potential mates. Lays clutches of 40 to 45 eggs in nest chamber within burrow. Omnivorous. Eats leaves, fruits, insects, and small vertebrates.

male

male

## Green Iguana (*Iguana iguana*)

79 in (200 cm). Largest lizard in Costa Rica. Conspicuous comblike spines on neck, back, and tail (absent in young); very large circular scale below ear opening; pendulous dewlap bears serrated edge. Young bright green; with age, color changes to gray, brown, greenish gray, or nearly black. Breeding males often show orange or yellow on head and body. Common; occurs on Caribbean and Pacific slopes to 1,640 ft (500 m). Diurnal; when sun is shining, emerges to bask on exposed branches. Prefers to perch on branches overhanging water; large individuals readily dive into water from great heights to escape danger. Adults descend to ground only to lay eggs; female lays up to 70 eggs during the dry season. Young mostly terrestrial; often form large social groups; feed on insects. Adults are mostly vegetarian, but would be unable to digest plant matter without aid of symbiotic microorganisms in the digestive tract.

## Green Spiny Lizard (*Sceloporus malachiticus*)

7.1 in (18 cm). Medium size. Robust body; blunt, short-snouted head; spine-tipped scales give overall spiky appearance. Bright green, yellowish green, olive, or brown. Capable of considerable color change in response to environmental cues and may darken to almost completely black. Male with bright-turquoise (or blue) tail and throat; black collar; on white belly, pair of blue patches (with black borders). Female and young usually brown with 9 or 10 pairs of dark blotches. Abundant; occurs on Caribbean and Pacific slopes between 1,970 and 12,470 ft (600 and 3,800 m). Found in cloud forests (on trees), high-elevation pastures (on rocks, logs, and fence posts), and in city parks and gardens. Conspicuous—frequently basking in exposed spots—but also wary, darting into crevices at least sign of disturbance. Breeds year-round at high elevations, but perhaps seasonally at lower elevations. Bears 6 to 12 live young.

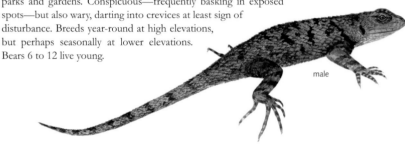

male

## Brown Spiny Lizard (*Sceloporus variabilis*)

6.7 in (17 cm). Medium size. Robust body; short-snouted head; spine-tipped scales. Brown overall, with a pair of pale stripes that run from behind the eye to the base of the tail. Belly and other undersurfaces are whitish with a pink to red hue; during the breeding season, male has a pair of red patches (with blue outline) on belly. Locally common; occurs only in the hot, dry N.W. Pacific, to 2,620 ft (800 m). Prefers fields, roadsides, beaches, and other open areas. Generally seen running across open ground or perched on logs, rocks, tree-trunk bases, or fence posts. Diurnal; active on hot, sunny days. Eats insects.

# Green Spiny Lizards

Male Green Spiny Lizard (*Sceloporus malachiticus*).

The male Green Spiny Lizard uses the brilliant colors of its throat and chest to display to females and to ward off other males in territorial disputes. But these very same displays can also draw unwanted attention from predators. When danger threatens, the Green Spiny Lizard presses throat and chest against the object on which it rests, thereby concealing its bright colors.

This species is one of the few lizards that is common in highland regions of Costa Rica. There it seeks the driest and warmest spots it can find. When the sun rises, it positions itself against a dark surface—often a log, fence post, or tree trunk—so as to maximize heat intake. Highland populations are darker in coloration than lowland populations, which also aids in heat absorption.

## Green Tree Anole (*Norops biporcatus*)

13 in (33 cm). Large with short and robust limbs. Fingers and toes with partially expanded pads. Bright leaf-green coloration, occasionally with pale-blue spots. When stressed, changes to mottled dark brown with white and black spots. Both male and female have small dewlap (not always visible in female) with red margin, blue center, and white base. Occurs on Caribbean and Pacific slopes to 3,940 ft (1,200 m). Found in relatively undisturbed humid forests. Usually alone or in pairs. Strictly diurnal; at night, sleeps on exposed branches or on top of palm fronds. Arboreal; generally observed within human reach but also ascends to the canopy. Overpowers relatively large invertebrates using strong jaws; occasionally eats small anoles. Gapes and attempts to bite when threatened. May vocalize when handled.

## Pug-nosed Anole (*Norops capito*)

9.8 in (25 cm). Medium size. Short head; pug-nosed snout. Very long tail has clublike tip. Fingers and toes partially expanded to form pads. Coloration variable. Male with small dewlap (greenish-yellow or olive). Occurs on Caribbean slope and in S. Pacific, to 3,940 ft (1,200 m). Found on the forest floor or perched low on a trunk. A skillful climber. Effective camouflage and habit of pressing its body against tree trunks make spotting this lizard a challenge. When approached too closely, adults may run for cover on hind limbs. Powerful predator, capable of overpowering relatively large insects and other invertebrates; also known to prey on smaller anoles.

male

## Pacific Anole (*Norops cupreus*)

6.3 in (16 cm). Smallish. Relatively slender. Expanded pads on toes and
fingers increase ability to climb vertical surfaces. Dull brown or grayish
brown. Male dewlap very large; base reddish orange or yellow-orange, margin
pink. Female lacks dewlap but sometimes has a pink-and-orange spot on the
throat. Abundant and conspicuous; occurs on Pacific coast (as far south as
Dominical) and, locally, in the Central Valley, to 4,595 ft (1,400 m). Found in
dry habitats in disturbed areas, which it seems to prefer over forests. Active
year-round in leaf litter, bushes, and trees, and on buildings. Feeds on a wide
variety of invertebrates.

## Ground Anole (*Norops humilis*)

4.3 in (11 cm). Small, stocky. Fingers and toes partially expanded into pads.
Grayish brown to chocolate brown; back marked with chevron- or diamond-
shaped spots. On top of head, there is often a dark band that connects the
eyes. Male has small red dewlap with yellow margin. Tube-like pockets in
armpits, often filled with mites, are unique to this species. Abundant; occurs
on Caribbean slope and, locally, on the Pacific slope (where it is most common
in the S. region), to 4,920 ft (1,500 m). Found in leaf litter in a variety of
habitats, both pristine and disturbed. Mostly terrestrial, although adult male
performs territorial dewlap displays from low perches. Eats insects. Female
lays a single egg, but may do so every week. Young, sexually
mature at 6 months, rarely live beyond 2 years.

## Slender Anole (*Norops limifrons*)

5.9 in (15 cm). Small, very slender. Attenuated snout; long tail; fingers and toes with expanded pads. Tan, gray brown, olive brown, or reddish brown, with distinct broad black bands on the tail. Often with a pair of thin black lines along the middle of the back. Lips, chin, throat, and undersurfaces white. Very small dewlap on male is white with a yellow-orange spot at its base. Abundant; occurs on Caribbean slope and on central and southern Pacific slope, to 4,430 ft (1,350 m). Most common in dense secondary vegetation or at forest edges. Generally seen perched in a head-down position on low vegetation or tree trunks but also descends to leaf litter. Agile, seemingly always on the move. Forms loose male-female bond, and pairs generally travel through the forest in close proximity. Female produces a single egg every 7 to 11 days. *Illustration not to scale.*

## Stream Anole (*Norops oxylophus*)

9.4 in (24 cm). Medium size. Expanded pads on fingers and toes. Dark olive-brown with white undersurfaces and a distinct white stripe on each side of the body; there is often a series of white spots (with dark outline) above the white stripe. Male with an erectable low crest on back and tail; dewlap large, burnt orange. No dewlap in female. Locally common; occurs on Caribbean and Pacific slopes to 3,940 ft (1,200 m). Rarely found far from streams; usually perched on rocks or logs near water's edge. When approached, escapes into stream and may swim considerable distance under water. Often remains submerged, hiding in air pockets below rocks and logs. *Illustration not to scale.*

male

**Family TEIIDAE (Whiptailed Lizards).** Fast-moving terrestrial lizards with a stream-lined body, long, whiplike tail, and powerful limbs. Commonly seen basking in sun-exposed locations along roads, trails, plantations, and beaches, where it is often possible to approach quite closely before the lizards suddenly dash away.

## Central American Ameiva (*Ameiva festiva*)

13.4 in (34 cm). Medium size. Narrow head; pointed snout; long, slender tail. A cream (to pale-yellow) stripe runs from the tip of the snout to the base of the tail (stripe is blue on tail). Body olive, dark gray, brown, or almost black; young often with reddish markings. Large males with blue head and throat during mating season. Abundant; occurs on Caribbean slope and in S. Pacific to 3,940 ft (1,200 m). Predominantly inhabits tree-fall areas, road cuts, and forest edges. Basks in sun-exposed spots. Moves jerkily across forest floor, scanning for insect prey; frequently stops to dig in leaf litter with front legs. Female lays clutch of 1 to 4 leathery eggs in shallow nest, 3 or 4 times per year.

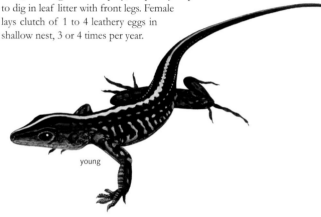

young

## Four-lined Ameiva (*Ameiva quadrilineata*)

11 in (28 cm). Medium size. Attenuated head; long, slender tail; four pale stripes run down the length of the body. Breeding males with orange or red on head and throat. Abundant; occurs on the Caribbean slope and on the Pacific slope (from the central coast south), to 3,280 ft (1,000 m). Inhabits forest edges, roadsides, plantations, and beach-front vegetation. Prefers hot, dry areas; on overcast days, remains inactive in burrow. Agile; moves with jerky motions between clumps of vegetation, probing leaf litter for small invertebrates. Buries clutches of 1 to 3 eggs in loose soil.

breeding male

**Family SCINCIDAE (Skinks).** There are three species in Costa Rica, each in a different genus. Skinks are smooth-scaled and shiny and have short limbs and a cylindrical body and tail. They move rapidly through leaf litter and dense vegetation, sometimes moving in snakelike undulations, without the use of their limbs.

## Litter Skink (*Sphenomorphus cherriei*)

6.7 in (17 cm). Small. Head short, bluntly rounded, and only slightly wider than the neck. Has moveable eyelids. Brown to bronze, with minute black and tan spots. A dark band, of varying length, runs from the tip of the snout along the side of the body. Abundant; occurs on the Caribbean slope and in the S. Pacific, to 3,610 ft (1,100 m). Found in leaf litter of humid forests; most common in open-edge situations. Usually glimpsed as it darts in and out of the leaf litter; rarely basks in sun-exposed spots. Tail used to store fat reserves and to maintain balance; may avoid predators by flipping its entire body 180 degrees while running, using the heavy tail for leverage. Most individuals have a regenerated tail. Throughout the year, female produces clutches of 1 to 3 eggs.

**Family GYMNOPHTHALMIDAE (Microteiid Lizards).** A predominantly South American family. The few species in Costa Rica are all quite different in appearance and biology.

## Golden Spectacled Lizard (*Gymnophthalmus speciosus*)

4.7 in (12 cm). Small. Lacks moveable eyelids; eyes covered with clear transparent spectacle. Ear openings clearly delineated. Only four fingers on each hand (five toes on the hind feet). Head and back are bronze to gold; tail bright brick-red, especially in young. Common and widespread; occurs on Caribbean and Pacific slopes to 3,940 ft (1,200 m). Found in gardens, plantations, road banks, and other disturbed areas. Often overlooked since it is secretive and usually hidden in leaf litter. Most active during hottest hours on sunny days; retreats into burrow or under rocks and logs if temperature drops. Each year female produces several small clutches of eggs.

**Family XANTUSIIDAE (Night Lizards).** A family of unusual looking lizards, characterized by snakelike, lidless eyes and pointed scales scattered over the body and tail. Despite common name, these lizards are mostly active at dusk and dawn.

## Yellow-spotted Night Lizard (*Lepidophyma flavimaculatum*)

9.4 in (24 cm). Medium size. Large, bluntly rounded head; body, tail, and limbs covered in small granular scales, interspersed with spikelike projections. Eyes lack moveable eyelids. Dark gray, dark brown, or black, with a pattern of distinct yellow, cream, or white rosettelike spots. Alternating dark and light bars on lips. Common but often in widely separated populations; occurs on Caribbean slope to 2,460 ft (750 m). (Similar-looking Reticulated Night Lizard, *Lepidophyma reticulatum*, occurs only in Pacific lowlands.) Very secretive. Basks at the entrance of burrow, often with only head protruding, and retreats rapidly when approached. Never ventures far from burrow. Many populations are all-female; capable of producing all-female offspring without fertilization (parthenogenesis). Live-bearing; produces litters of up to six young during rainy season.

**Family ANGUIDAE (Anguid Lizards).** A family of smooth-scaled lizards that generally have relatively small limbs and a cylindrical body and tail.

## Galliwasp (*Diploglossus monotropis*)

21 in (53 cm). Large, robust, with a fairly cylindrical body and tail. Unique color pattern of bright orange to red markings on the belly, flanks, and head. Occurs on the Caribbean slope in lowland rainforests, to 1,640 ft (500 m). Diurnal. Although large and colorful, this secretive lizard is a challenge to see. Most frequently observed as it moves through leaf litter—but flees rapidly when discovered. Little is known about its biology, but it is thought to be a specialized predator of land crabs. Because of its bright colors, many country folk mistakenly associate the nontoxic Galliwasp with the venomous coral snakes, a misconception enshrined in some of its local names: *madre de culebra* (mother of serpents), *madre coral* (mother of coral snakes), and *escorpión coral* (coral scorpion).

**Family BOIDAE (Boas).** This New World family of giant snakes includes such charismatic species as the Boa Constrictor and the anacondas of South America. Boids tend to be much bulkier than their Old World counterparts, the pythons, and large individuals are a formidable sight. Costa Rican boas have heat-sensitive organs in their lip scales (not visible externally) that help direct strikes toward any prey that is warm-blooded. Costa Rica is home to four species, in three genera. The Boa Constrictor is the most commonly seen.

## Boa Constrictor (*Boa constrictor*)

To 14.5 ft (445 cm), but generally under 6.5 ft (200 cm). Largest (and heaviest) snake in Costa Rica. Superficially viperlike but lacks triangular head and heat-sensitive pit between eye and nostril. Broad, dark eye stripe on sides of head. Spots, bars, and diamonds of various colors adorn back. Widespread; occurs countrywide to 4,290 ft (1,500 m). A habitat generalist: found in relatively undisturbed forest, cultivated fields, and areas near human settlements. Nocturnal and diurnal; terrestrial and arboreal. Hides in burrows or tree holes when inactive. Ambushes, constricts, and eats lizards, birds, and mammals to the size of coatis. Curved, needlelike teeth can inflict deep lacerations. Bears live young (up to 60 per litter).

**Family COLUBRIDAE (Colubrid Snakes).** In essence, this family is a collection of all the snakes that do not fit neatly into any of the other snake families. It is thus a large family—with nearly 90% of the world's snakes—and extremely diverse. Future research will undoubtedly lead to splitting it into more sensible, less unwieldy groupings. Colubrid snakes range from nonvenomous constrictors to venomous, rear-fanged species. Although no Costa Rican colubrids are known to be fatally venomous, many possess some type of venom. It is best, therefore, to avoid handling wild snakes.

## Mussurana (*Clelia clelia*)

7.8 ft (240 cm). Large, powerful. Cylindrical body covered with smooth, shiny scales. Vertically elliptical pupils. Fangs located at back of mouth. Adult uniform blue-black to dark gray; pale belly and throat. Subadult often has pale band across neck. Young have striking, bright-red body; black head; broad yellow band across nape. Common and widespread; occurs on Caribbean and Pacific slopes to 2,950 ft (900 m). Inhabits a variety of forest types and also disturbed habitats. Terrestrial. Primarily nocturnal but frequently active during the day. Eats other snakes, including venomous pitvipers (Mussurana is immune to pitviper venom). Uses venom and constriction to dispatch prey. When handled, does not generally attempt to bite, but effects of its venom on humans is largely unstudied, and large individuals should be treated with care. Egg-laying.

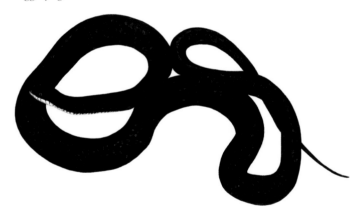

## Brown Litter Snake (*Coniophanes fissidens*)

31.5 in (80 cm). Small to medium size. Rounded head; thick body; long tail. Brown snake with faint stripes (sometimes bordered by a thin dark line or series of small dark spots); striped pattern more pronounced on tail, though many individuals lack at least part of the tail. White stripe behind each eye is usually bordered above by a dark-brown marking. Large eyes, round pupils, copper irises. Lip stippled with pale and dark patches. Common to abundant; occurs on Caribbean and Pacific slopes to 2,950 ft (900 m). Found in leaf litter of dense forests, secondary vegetation, and shaded plantations. Feeds on variety of insects, amphibians, and reptiles and their eggs. Fangs located at back of mouth. Mildly venomous; bite may cause local swelling, pain, and sensitivity. If the tail is restrained (by predator or otherwise), it can break off; tail does not regenerate. At the outset of the rainy season, female lays clutches of 1 to 7 eggs.

## Roadguarder (*Conophis lineatus*)

51 in (130 cm). Medium size. Cone-shaped head is scarcely wider than the neck. Cream to tan, with bold, dark stripes that run along length of body (some individuals are uniformly dark). Occurs on Pacific slope, principally in the N.W. Pacific but also in the Central Valley, to 3,610 ft (1,100 m). Prefers dry or semi-dry open habitats with low vegetation; commonly found along the side of roads, hence its common name. If pursued, often escapes down burrows; may spend considerable time underground. A very fast-moving predator of lizards, especially whiptailed lizards of the genera *Ameiva* and *Cnemidophorus*; also eats frogs, toads, snakes, small mammals, and bird eggs. Uses greatly enlarged rear fangs to inject venom into prey. In humans, bites can cause pain, local swelling, headache, vomiting, and discomfort; bites often bleed profusely due to anticoagulants in venom. Produces clutches of 2 to 6 eggs.

### False Coral Snake (*Erythrolamprus mimus*)

37 in (95 cm). Medium size. A coral snake mimic. Bold black, white, and red rings encircle the body, but pattern sometimes faint on belly. Red markings may be present on the tail, unlike in true coral snakes (genus *Micrurus*). Fairly uncommon; occurs on Caribbean slope and in S. Pacific to 3,940 ft (1,200 m). Found in humid forests. Biology similar to that of coral snakes: a secretive inhabitant of the forest floor; most active under low-light conditions; feeds mainly on snakes, but occasionally eats lizards. Fangs located at back of mouth; its mild venom can cause local pain and swelling in humans.

### Gray Earth Snake (*Geophis brachycephalus*)

18 in (46 cm). Small. Smooth, bullet-shaped head; short tail; iridescent scales. Tiny eyes with round pupil and black iris. Pale belly. Three distinct color patterns known: black with red blotches on the posterior half of the body that are sometimes fused to form crossbands (as in illustration); uniform gray or black; gray with a pair of red, lateral stripes. Young have a distinct white collar. Common; occurs in the Central and northern Talamanca mountain ranges, the Central Valley, and in scattered locations in the N. Caribbean and S. Pacific, to 6,890 ft (2,100 m). Inhabits a variety of habitats, from pristine rainforests to garbage piles in urban areas. Generally found when turning over objects; occasionally seen in nests of leaf-cutter ants. Ground-dwelling; uses insect tunnels to travel; may surface on rainy nights. Feeds on insect larvae, worms, and other soft-bodied invertebrates. Deposits 3 to 6 eggs in soil; in areas without a pronounced dry season, it may reproduce year-round.

## Brown Blunt-headed Vine Snake (*Imantodes cenchoa*)

49 in (125 cm). Medium size. Exceedingly thin and elongate, with distinctly compressed body. Wide head. Large bulging eyes; gold irises; vertically elliptical pupils. Dark, saddlelike blotches (reddish brown to chocolate brown) on silvery-gray, tan, or pale-brown background. Common and widespread; occurs on Caribbean and Pacific slopes to 4,920 ft (1,500 m); absent from dry N.W. Pacific. Nocturnal. During the day, hides under loose bark, in bromeliads, or on low vegetation. Arboreal, moving through vegetation from near ground level to high in the canopy. Can horizontally extend itself up to half the length of its body to bridge gaps between branches. Preys on sleeping anole lizards (genus *Norops*), using grooved rear fangs to administer a weak but paralyzing venom (harmless to humans). Egg-laying.

## Tropical Milk Snake (*Lampropeltis triangulum*)

79 in (200 cm). Individuals range from medium size to very large. Rounded head. Small eyes have round pupil and black iris. Red, white, and black rings encircle its body, creating a pattern similar to that of venomous tricolor coral snakes (genus *Micrurus*), but with the rings in a different sequence. Red rings bordered on each side by a black ring (red rings bordered on each side by a yellow or white ring in coral snakes). Amount of black pigment may increase with age in some populations; some adults are almost entirely black. Widespread but generally occurs in small numbers in any given area; found throughout the country to almost 8,200 ft (2,500 m). A habitat generalist; inhabits both dry scrub-forests and cool, wet cloud forests. Generally terrestrial. Nocturnal or crepuscular in many habitats, but in dense, shaded forests it tends be more diurnal. Non-venomous. A powerful constrictor that feeds on mammals, birds, reptiles, and eggs. Egg-laying.

### Northern Cat-eyed Snake (*Leptodeira septentrionalis*)

39.5 in (100 cm). Medium size. Head distinctly wider than neck. Large, bulging eyes with vertical, elliptical pupil and reddish iris; a dark band or spot present behind the eye. Tan, light brown or reddish brown, with dark-brown or black blotches that in some individuals fuse to form a zigzag band. Young with distinct white band across nape. Abundant; occurs on Caribbean and Pacific slopes to 3,940 ft (1,200 m). Found in a wide variety of habitats, but particularly common in vegetation at edges of lowland ponds where amphibians breed. At night, this snake hunts for frogs and their eggs; renowned for feeding on the arboreal eggs of Red-eyed Tree Frogs (*Agalychnis callidryas*). Swallows eggs and small frogs alive, but kills larger frogs with mild venom administered through enlarged rear fangs. May flatten head and spread jaws when threatened, lending it a menacing, viperlike appearance (although it rarely bites). Produces clutches of 6 to 13 eggs.

### Green Parrot Snake (*Leptophis ahaetulla*)

86.5 in (220 cm). Very large. Relatively slender with a long tail (often incomplete). Leaf green above, pale green on belly and lip. A short black stripe behind each eye; often has a black mark between eye and nostril. Eyes large with round pupil and yellow iris. (Similar to Satiny Parrot Snake, *Leptophis depressirostris*, though that snake never has a black mark between eye and nostril.) Widespread; occurs on Caribbean and Pacific slopes to 4,590 ft (1,400 m); absent from Nicoya Peninsula and driest areas of Guanacaste Province. Found in humid lowland forests. Diurnal. Both terrestrial and arboreal. Very agile and fast-moving; usually seen crawling through low vegetation hunting frogs, lizards, and other vertebrates. Sleeps coiled on high branches. Relies on an impressive open-mouth defensive display to ward off predators. Bites if handled. Fangs located at rear of mouth; saliva may contain anticoagulants because bite wounds bleed profusely. Egg-laying.

## Salmon-bellied Racer (*Mastigodryas melanolomus*)

55 in (140 cm). Large. Bold checkerboard pattern on young gradually darkens with age to form adult coloration of brown background marked with 1 or 2 faint stripes (cream to gray) along the sides. On most individuals, an irregular dark stripe extends from nostril beyond eye. Common; occurs throughout country to 5,580 ft (1,700 m). Regularly encountered in relatively pristine forests and mature secondary forests of humid lowlands; less abundant in dry forests of foothills. Nonvenomous. Eats lizards but also other small vertebrates. Strictly diurnal. Both terrestrial and arboreal; active on forest floor and in low vegetation but sleeps coiled on relatively high branches. Produces clutches of 1 to 9 eggs.

adult

## Red Coffee Snake (*Ninia sebae*)

13.8 in (35 cm). Tiny. Small head and slender body. Eyes small with round pupil and black iris. Coloration brick-red to red (sometimes marked with dark spots). Broad yellow collar bears a large, saddlelike black marking. Upper lips often yellowish cream. Belly is white. Abundant; occurs on Caribbean and Pacific slopes to 3,610 ft (1,100 m). Found in leaf litter of plantations, gardens, and secondary growth (uncommon in undisturbed habitats), often in moist spots under logs or debris. When threatened, flattens its body to an extreme degree or frantically thrashes about, but it is completely harmless. Local name is *vibora de sangre* (blood viper), and some people erroneously believe it to be fatally venomous. Produces 2 to 4 eggs per clutch.

## Brown Vine Snake (*Oxybelis aeneus*)

67 in (170 cm). Medium to large size. Extremely slender; very long tail; elongated head. Grayish brown, yellowish brown, or tan (sometimes peppered with tiny spots). Dark eye stripe extends onto neck. Eyes face forward to enable binocular vision. Lining of mouth—which is strikingly purple-black—is displayed in defensive gape. Common and widespread; occurs on Caribbean and Pacific slopes to 4,270 ft (1,300 m); absent in dry N.W. Pacific. Often remains motionless, typically with tongue extended. When wind stirs surrounding vegetation, the snake gently sways its body to mimic the movement of the plants. Easy to overlook. Diurnal; hunts lizards, frogs, insects, and small birds and mammals. Prey is immobilized with mild venom administered through enlarged, grooved rear fangs. Venom may cause local swelling and blistering in humans. Egg-laying.

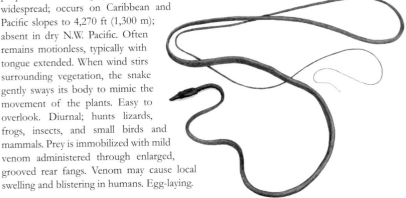

## Bird-eating Snake (*Pseustes poecilonotus*)

98 in (250 cm). Very large. Robust, with long, whiplike tail. Highly variable, but adults generally dark (blue-gray, gray, green, olive, or brown), often with contrasting colored transverse stripes or mottling; belly gray, cream, or yellow. Young cream or yellowish, with a pattern of dark V- or crescent-shaped crossbands. Common; occurs on Caribbean and Pacific slopes to 4,270 ft (1,300 m); absent from dry N.W. Pacific. An agile climber; generally seen on low vegetation in forests, although it can be found on the ground as well. Diurnal. Diet consists primarily of birds and their eggs, but occasionally eats bats. When threatened, flattens head, puffs up neck and front of body, gapes, hisses, then strikes. Nonvenomous.

adult

## Red-bellied Litter Snake (*Rhadinaea decorata*)

18.5 in (47 cm). Small. Dark brown; a broad bronze stripe runs down the length of the back. A pair of conspicuous dark-rimmed white spots present behind each eye and on the nape. Throat and upper lip white or cream. Caribbean populations with a distinctive red belly (belly sometimes white in S. Pacific populations). Common; occurs on the Caribbean slope and in the S. Pacific, to 3,940 ft (1,200 m). Found in the leaf litter of humid, relatively undisturbed habitats, but occasionally appears in secondary growth. Diurnal. Hunts small frogs, tadpoles, frog eggs, and, rarely, small lizards. Excretes foul-smelling musk when captured. May break off tail to elude predators; many adults lack part of the long tail, attesting to the effectiveness of this escape strategy.

## Common Snaileater (*Sibon nebulatus*)

32.5 in (83 cm). Smallish. Bluntly rounded head; large eyes; prehensile tail. Color pattern variable; generally grayish to dark brown with irregular dark bands that are outlined in white. Common; occurs on the Caribbean slope and in the S. Pacific, to 3,610 ft (1,100 m). Also occurs in a few isolated locations in the Central Valley. Favors humid and wet forests. Nocturnal and arboreal. Has modified teeth and jaws suited for extracting snails from their shell; feeds exclusively on snails, slugs, and other soft-bodied prey. To defend itself, sometimes flattens head and spreads jaws to create a viperlike, triangular head shape. Nonvenomous, harmless. Produces clutches of up to 9 eggs.

# Hard to Swallow

Black Halloween Snake (*Urotheca euryzona*) eating a rain frog (*Craugastor* sp.).

Contrary to the popular notion, snakes do not "dislocate" their jaws to swallow large prey. Instead, the lower jaw—a set of separate mobile bones—can expand outward to accommodate larger animals. These same bones are set in motion to assist in moving the prey animal down the digestive tract.

Snakes usually consume prey head-first. That way, as the animal is ingested its limbs fold flat against its body; a second advantage is that any fur or scales on its body does not obstruct passage. Snakes determine the location of the prey's head based on the "grain" of its scales, hairs, or feathers. In the case of this smooth-skinned rain frog, however, the snake apparently lacked sufficient information to tell which end was up. (Not described within the species accounts.)

189

### Tiger Rat Snake (*Spilotes pullatus*)

98 in (250 cm). Very large. Long whiplike tail. Eyes large with a round pupil and black iris. Characteristic "tiger stripe" pattern of cream (to yellow) streaks, bars, or spots on a black background; often tail and posterior part of body completely blue-black. Common; occurs on Caribbean and Pacific slopes to 4,590 ft (1,400 m); absent from dry N.W. Pacific. Found in a variety of habitats, including dense rainforest, swamps, and scrub vegetation. Diurnal; often spotted at stream banks, trails, or forest edges. Both terrestrial and arboreal (readily ventures into high canopy). Hunts small mammals, lizards, birds, and bird eggs. Employs impressive defensive display, rearing up with neck inflated as it rattles tail tip against dry leaves, creating an audible buzz. Stands its ground; hisses loudly and strikes repeatedly when threatened. Nonvenomous. Produces clutches of 7 to 10 eggs.

### False Fer-de-Lance (*Xenodon rabdocephalus*)

39.5 in (100 cm). Medium size. Robust body; triangular head; short tail. Brown with hourglass-shaped, dark brown blotches that are often outlined in white. Top of head with dark brown arrow-head outlined by tan or silver-gray. Dark-brown eye stripe from tip of snout to corner of mouth. Round pupils, smooth scales, and absence of heat-sensitive pits differentiate it from similar Fer-de-Lance (*Bothrops asper*). Occurs on Caribbean and Pacific slopes to 3,280 ft (1,000 m). Fairly common on Caribbean slope and in S. Pacific; rare in N.W. Pacific and Central Valley. Favors humid areas; often found along streams and ponds in lowland forests. Eats toads and, less frequently, large frogs. Has greatly enlarged teeth at back of mouth that it uses to pierce and deflate toads that inflate their body in defense. Appears immune to amphibian skin toxins. Irascible temper; hisses and strikes when threatened. Rear fangs may cause painful lacerations and profuse bleeding.

**Family ELAPIDAE (Coral Snakes and relatives).** This family includes such charismatic venomous species as the cobras, mambas, kraits, coral snakes, and sea snakes. Four species of terrestrial coral snake occur in Costa Rica. Three have the characteristic color pattern of red, yellow, and black rings; the fourth generally has black and orange bands. A single sea snake species visits Costa Rica's shores. Some authorities assign all sea snakes to a family of their own, but traditionally they are placed in a subfamily of Elapidae. The four coral snakes and the sea snake that inhabit Costa Rica are all dangerously venomous and can cause human fatalities. Their venom consists of neurotoxins that paralyze heart and lung muscles.

## Bicolored Coral Snake (*Micrurus multifasciatus*)

47 in (120 cm). Medium size. Head is scarcely wider than the neck; cylindrical body; very short tail. Eyes diminutive, with round pupil and black iris. Generally has orange and black rings, but in some individuals orange is replaced by pink or white. Occurs on Caribbean slope to 3,940 ft (1,200 m). Usually seen crawling through leaf litter in deeply shaded zones of undisturbed rainforest, especially during first hours of the day and in late afternoon. Mainly eats snakes and other reptiles. When threatened, moves erratically, hides head under body coils, and presents coiled tail tip to predators; bites readily. Venom contains strong nerve toxins. Bites are very rare but should be considered extremely serious.

defensive
tail display

## Central American Coral Snake (*Micrurus nigrocinctus*)

45 in (115 cm). Medium size. Head is scarcely wider than the neck; cylindrical body; very short tail. Snout bluntly rounded. Eyes are very small (with round pupil and black iris) and are often difficult to discern as they are located within black marking. Has typical coral snake pattern of black-yellow-red-yellow-black body rings that completely encircle the body (no red on head and tail). In Pacific slope populations, the broad yellow rings are replaced by narrow pale-yellow or white rings. The most commonly encountered coral snake in Costa Rica; occurs on Caribbean and Pacific slopes (including Central Valley), to 4,920 ft (1,500 m). Found in a variety of habitats, including dry rocky areas, marshes, rainforests, and even cultivated fields and gardens. Secretive; inhabits soil, leaf litter, interior of logs or stumps, and rock crevices. Both diurnal and nocturnal. Preys on snakes, caecilians, and elongate lizards such as skinks. Deposits small clutch of elongated eggs on the forest floor, either in leaf litter or soil. Eggs hatch after 2 to 3 months, generally at start of rainy season.

Caribbean
slope
population

## Yellow-bellied Sea Snake (*Pelamis platurus*)

32.5 in (83 cm). Medium size. Only sea snake in Central America. Has oarlike, compressed tail. Top of head, back, and tail covered with broad black band; undersurfaces generally bright yellow but belly brown in some individuals. Occasionally entirely yellow. Occurs in Pacific Ocean. Pelagic; spends its entire life drifting on ocean currents. Often found in large numbers floating amid flotsam. Individuals sometimes wash onto Pacific beaches after storms and lack muscle power to return to water; when exposed to sun they overheat and die. An agile swimmer and diver; can remain submerged up to 90 minutes. Sheds skin by coiling body in knots and using friction to remove old skin. Reproduces at sea; mates while floating near the water surface and gives birth to 5 or 6 live young after gestation period of at least 5 months. Kills fish with potent venom. Humans are rarely bitten, and symptoms are normally mild, but fatalities have been recorded. Does not frequent shallow waters and poses no threat to swimmers.

**Family VIPERIDAE (Vipers).** The 15 vipers that occur in Costa Rica are all in the subfamily Crotalinae (Pitvipers). Distinguishing features are a heat-sensitive pit on each side of the head between the eye and the nostril; a triangular head and a stocky body; cryptic coloration that allows the snake to ambush passing prey; and long, hollow fangs normally folded back against the roof of the mouth but erected during a strike. The needlelike fangs deliver a venom that is a cocktail of toxins and enzymes that cause both extensive tissue damage to the area of the bite and, often, permanent damage to internal organs. All species found in Costa Rica are capable of delivering a potentially dangerous bite, and a number of human fatalities are recorded every year in the country.

## Jumping Viper (*Atropoides mexicanus*)

37 in (95 cm). Medium size. Extremely robust; broad head; short tail. Dark stripe between eye and corner of mouth. Underside of head and throat usually white or cream. Top of head and back with prominently keeled scales. Gray, tan, or light brown, with dark-brown or black rounded spots on the middle of the back; spots sometimes fuse to form a zigzag band. Row of similarly colored blotches on each side of body. Young have bright-yellow tail tip that may be used as a lure. Occurs in isolated populations on Caribbean and Pacific slopes, to 4,590 ft (1,400 m). Found on forest floor of humid, relatively undisturbed forests. Mainly a diurnal, ambush hunter; prefers rodent prey but young also eat small lizards, frogs, and invertebrates. Prey killed with relatively mild venom, delivered with a lightning strike (when striking does not jump off the ground as is commonly believed). Live-bearing; may produce litters of over 30 young. (Formerly *Atropoides nummifer*.)

## Eyelash Pitviper (*Bothriechis schlegelii*)

19.7 in (50 cm). Smallish. Slender body; prehensile tail; spinelike scales over the eyes create the effect of eyelashes. Coloration highly variable, with different color forms sometimes appearing even within the same litter. Usually green, olive green, brown, or grayish brown, with cryptic blotches, spots, or crossbands. Another common, yet very distinct form—entirely bright yellow—goes by local name of *oropel*. Small young have cream or yellow tail tip that they sometimes use as a lure to attract prey. Common to abundant; occurs on Caribbean and Pacific slopes to 4,920 ft (1,500 m). Prefers moist forests, where it is often found at forest edges and along trails and streams; readily ascends into the high canopy. Nocturnal and arboreal. Spends daylight hours coiled on tree roots, branches, or bromeliads. Eats small vertebrates caught in ambush. Venom potent; snakebite incidents are fairly common as it is often coiled on vegetation in plantations and near trails. Live-bearing; produces litters of up to 19 young.

## Hognosed Viper (*Porthidium nasutum*)

Male 18 in (46 cm); female 25 in (63 cm). Small to medium size. Stocky. Distinctly upturned snout bears proboscis-like extension. Brown, reddish brown, or gray; dorsal pattern of obscure, dark blotches. Tail greenish in young snakes but changes to cream with bold dark splotches in adults; young may use colorful tail tip to lure small frogs within striking distance. (Similar White-tailed Hognosed Viper, *Porthidium porrasi*, is restricted to lowlands of Osa Peninsula and Golfo Dulce region.) Common; occurs on Caribbean slope to 2,950 ft (900 m). Terrestrial; inhabits leaf litter of humid and wet forests. Most frequently seen on trails or near buttress roots of large trees but often goes unnoticed due to excellent camouflage. Sometimes remains in same ambush site for several weeks awaiting suitable prey. Adult feeds on relatively large mammals, frogs, and lizards.

# A Terrible Beauty

Eyelash Pitviper (*Bothriechis schlegelii*).

The Eyelash Pitviper is responsible for a large number of snake bites in Costa Rica and ranks among the country's most dangerous snakes, a telling claim in a place with at least 20 species of venomous snake. All pitvipers have long, hinged fangs on the upper jaw that are used to inject venom deep into prey. This venom is a cocktail of sorts, containing tissue-destroying hemotoxins, digestive enzymes, and, in some species, neurotoxins that affect the nervous system. Such compounds not only immobilize (and kill) prey, they also aid in digestion, speeding up the time it takes for the snake to digest its meal and resume normal activity; intake of large prey can immobilize a snake for a period of time, during which it is highly susceptible to predators.

An arboreal creature, the Eyelash Pitviper is often found on the upper surface of leaves and branches, and occasionally on a vertical tree trunk (a very good reason to check before grabbing hold of a tree). Young Eyelash Pitvipers tend to occur closer to the ground than adults, a fact perhaps explained by the different diets of small and large pitvipers. Smaller individuals generally eat frogs and lizards, which occur more frequently on or near the ground, whereas larger (adult) snakes catch the birds and small mammals that inhabit treetops.

## Fer-de-Lance (*Bothrops asper*)

97 in (246 cm). Very large. Spanish common name is *terciopelo* (velvet), in reference to its velvety sheen. Dark-brown body bears cream or white chevrons and dark triangles (that sometimes fuse to form hourglass markings). Dark stripe between eye and corner of mouth. Young with yellowish tail tip. Widespread and abundant; occurs on Caribbean and Pacific slopes to 3,940 ft (1,200 m); absent from Nicoya Peninsula and dry regions of N.W. Pacific. Often found in dense vegetation, near streams, rivers, and other edge situations. Exceedingly common in agricultural areas where crops attract rodents. Terrestrial, although young snakes sometimes crawl into low vegetation. During the day, remains coiled (and concealed) in vegetation; at night, hunts frogs, lizards, and even mammals and birds. Agile, very fast, and excitable; possesses large venom glands that contain potent venom. Very dangerous; responsible for multiple human deaths each year. Live-bearing and extremely fecund; litter size may exceed 80 babies.

juvenile

## Neotropical Rattlesnake (*Crotalus durissus*)

71 in (180 cm). Large. The only snake in Costa Rica with a rattle. Has a distinct ridge along the back. Tan, yellowish brown, or grayish tan; two dark stripes on head extend onto neck; body bears a series of diamond-shaped, dark blotches. Common; occurs on Pacific slope, in Guanacaste Province, Nicoya Peninsula, and (historically at least) in the Central Valley, to 5,250 ft (1,600 m). Found in open dry forest and scrubland. Terrestrial and nocturnal. At night, actively hunts mammals and reptiles; rests in a coiled position during the day. Has a formidable defensive display: raises head off ground and hisses loudly, all the while producing a buzzing sound with its rattle. Quick to strike, and bite incidents are not uncommon. Rattle grows with age (young are born with only a single button on the tail tip). Live-bearing; gives birth to up to 30 young at the start of the rainy season, when prey is abundant. Males employ ritualized combat in territorial and courtship behavior.

## Central American Bushmaster (*Lachesis stenophrys*)

To 142 in (360 cm), but generally under 79 in (200 cm). Extremely large. Grayish brown, tan, or reddish brown, with a unique pattern of black, triangular shapes. Head with a broad, dark stripe between eye and corner of the mouth, below which there is usually a cream or pale-yellow patch. Occurs on Caribbean slope to 3,280 ft (1,000 m). (The Black-headed Bushmaster, *Lachesis melanocephala*, is endemic to the Osa Peninsula region.) Prefers undisturbed habitat in humid and wet lowland forests. Shy, secretive, and rarely encountered. Despite its fearsome reputation, it generally displays a relatively mild disposition. Eats Spiny Rats and other small mammals. Possesses very potent venom, and bites should be treated immediately. The only New World pitviper that lays eggs; unsubstantiated reports suggest that female defends nest site.

**Some 350 million** years ago, amphibians were the first vertebrate animals to successfully colonize land, and today there are about 6,500 species on the planet. Costa Rica is home to roughly 190 species divided into three orders: caecilians, salamanders, and anurans (frogs and toads).

Tropical amphibians display a dazzling variety in body shape, size, coloration, behavior, and biology. Caecilians, found exclusively in the tropics, are limbless, eyeless creatures that spend the majority of their time underground. They are the only amphibians in Costa Rica that produce live young.

Although tropical salamanders may resemble those found in temperate climates, they can be quite different in habits and biology. Many Costa Rican species, for example, are arboreal, inhabiting moss cushions and bromeliads in the rainforest canopy. All Costa Rican salamanders are lungless—they breathe through their thin skin. Their eggs, deposited not in water but on land, develop into tiny salamanders without passing through a tadpole stage.

Costa Rica is a great place to see frogs and toads, and many of them are stunning, fascinating creatures. The bright-red coloration of the Strawberry Poison Dart Frog warns potential predators of its toxicity. Particularly fascinating is the Horned Marsupial Frog, which carries its developing eggs in a pouch on its back. The Mexican Burrowing Toad remains underground almost year-round, emerging only to breed, during a brief period of time. When threatened, the Milk Frog excretes a toxic, gluelike substance from its skin. The gaudy Red-eyed Tree Frog, whose likeness appears on shirts, posters, and postcards, is a perennial favorite, as are glass frogs and the ubiquitous Giant Toad.

Most amphibians are nocturnal, and are best found with the aid of a powerful flashlight. Caecilians and tropical salamanders are elusive and rarely seen, but frogs and toads occur in almost any type of habitat and are easy to spot. During the breeding season male frogs of many species become highly vocal. It is then possible to locate calling individuals by sound. Lowland ponds on warm, rainy nights are an ideal spot to see large numbers of frogs.

Green-and-Black Poison Dart Frog (*Dendrobates auratus*). This gemlike frog can sometimes be located by listening for soft, repeated buzz calls emanating from leaf litter at the base of large tree buttresses in Carara, Corcovado, and other national parks.

## Order Gymnophiona (Caecilians)

There are seven species in Costa Rica, all in the family Caeciliidae. Caecilians are relatively featureless and superficially resemble large worms. Limbless, with very primitive eyes. Main sensory organs are short, moveable tentacles near eye sockets; these detect chemical cues and serve to locate prey or mates. Fossorial, usually underground. Infrequently seen but occasionally found under logs or other surface objects; sometimes flushed from burrows during heavy rains.

## Purple Caecilian (*Gymnopis multiplicata*)

19 in (48 cm). Long, stout. Eye sockets, covered by bone, are visible only as a pale area on each side of head. Tentacle located at the front edge of each eye socket (between the eye and nostril in other Costa Rican caecilians). Uniform gray to purplish gray with annular skin folds that are slightly paler in color. Widespread in the lowlands of both coasts to 4,590 ft (1,400 m), but absent from the drier regions of the N.W. Pacific. Gives birth to live young that feed during development on a nutritious, milklike substance secreted inside the female's body. Eats invertebrates and their larvae. *Illustration not to scale.*

## Order Caudata (Salamanders)

There are 44 species reported from Costa Rica, with several more known but still not described. All Costa Rican species belong to the family Plethodontidae, a group of lungless salamanders that absorb oxygen predominantly through the skin and the lining of the mouth. All Costa Rican members of this family lack an aquatic larval stage; the metamorphosis from larva to salamander takes place within eggs that are deposited in moist sites. Most species are restricted to humid forests at middle and high elevations, though some inhabit lowlands. Can "shed" tail to escape capture, but a fully functional replacement quickly regrows. A groove below each nostril (nasolabial groove) helps detect chemical cues that are used to locate prey, predators, and mates. Tropical salamanders are small, slow-moving, and strictly nocturnal. These secretive animals are poorly known, and very rarely seen.

In the species descriptions that follow, sizes represent the combined length of body and tail.

## Ridge-headed Salamander (*Bolitoglossa colonnea*)

4.3 in (11 cm). Relatively slender; medium size. A unique fleshy ridge connects the eyes. Hands and feet are extensively webbed. Color pattern and skin texture change from day to night. During the day, skin is dark with pale mottling; ridges show along the length of the body (resembles a small stick). At night, skin turns a uniform salmon, grayish, or tan, sometimes with small dark specks; body is smooth. Fairly common; occurs locally on Caribbean slope and in S. Pacific to 4,070 ft (1,240 m). Found in humid forests. Most active on warm, humid nights, when it explores low vegetation. A skillful climber. Generally moves in an extremely slow, deliberate manner; when disturbed, twitches erratically and sometimes launches itself off vegetation using heavy tail for leverage. With tail curled tightly around its body, sleeps hidden inside rolled-up dead leaves in leaf litter. Occasionally encountered in pairs, with both individuals in close proximity.

daytime appearance

### Red-legged Mountain Salamander (*Bolitoglossa pesrubra*)

5.1 in (13 cm). Robust. Hands and feet are moderately webbed. Extremely variable in color. Generally dark brown, gray, slate, or black (with paler mottling or spots overall). Limbs are red, orange, or ocher; throat is reddish or pink. Restricted to Talamanca Mountain Range, between 6,140 and 11,880 ft (1,870 and 3,620 m). Found in pristine páramo habitat but also disturbed areas (road sides, power-line cuts). Formerly abundant but populations have declined precipitously in last two decades for unknown reasons. During the day, hides in leaf litter, inside or under logs, in clumps of moss, and in bromeliads. On humid nights, explores crevices and vegetation for insect prey. Female deposits a clutch of 13 to 38 eggs in a moist spot (under a log or rock), then curls her body and tail tightly around the eggs. Female's skin secretions have antimicrobial properties and the close contact of her skin also helps to prevent the egg clutch from drying out.

### Pacific Worm Salamander (*Oedipina pacificensis*)

6.9 in (17.5 cm). Exceedingly slender, with a very long tail. Grayish black; tiny hands and feet; short limbs. Occasionally has irregular whitish blotch on top of the head and pale irregular spots on the limbs. Elusive but relatively common. Occurs in lowlands of S. Pacific, below 2,390 ft (730 m). Relatively tolerant of changes to habitat and may persist in plantations and other agricultural areas with dense vegetation. Favors humid regions. Forages in leaf litter or on low vegetation at night, but is most frequently found during the day, either under decaying logs or in leaf-litter accumulations between buttress-roots of large trees. Thirteen additional species of worm salamander occur in Costa Rica, all very similar in appearance. Although recognizable as a group, individual species are difficult to identify.

# Cool Climate Amphibians

Bark-colored Salamander (*Bolitoglossa lignicolor*).

Salamanders as a rule prefer temperate climates, so it is understandable why most salamander species in tropical Costa Rica occur at middle and high elevations. All species in the country are members of the family Plethodontidae, lungless salamanders that breathe through the skin and come equipped with a pair of grooves running from nostril to mouth that are used for chemoreception.

The high humidity that prevails in the habitat of the Bark-colored Salamander keeps its skin moist, which is necessary for the absorption of oxygen. But humidity plays another vital role in the life of this salamander and others of its family, for without a moist habitat it would be rendered "blind." By tapping the ground with its nose, it collects water-borne chemicals with the grooves on its snout. Organs in the nostrils, in turn, *interpret* these chemicals and assist in detecting mates, predators, and prey. The organs are amazingly sensitive. In laboratory experiments, these salamanders were able to determine that a snake had previously crossed their path—and to identify the direction in which it was headed! (Not described within the species accounts.)

**Family RHINOPHRYNIDAE (Mexican Burrowing Toad).** This family contains just a single species, the primitive Mexican Burrowing Toad. Unlike all other amphibians in Costa Rica, it lacks ribs, has only four toes (five in all others), and its tongue is attached to the back of the mouth (front of the mouth in other frogs). Sticks its calloused snout into a termite tunnel or ant trail and catches prey with its specialized tongue.

## Mexican Burrowing Toad (*Rhinophrynus dorsalis*)

3.5 in (8.9 cm). Large. Odd, primitive appearance. Small, pointed head, bloated body, and stocky arms and legs. Chocolate brown, purplish brown, or dark gray, with a yellow or reddish-orange stripe down middle of back. Common but seen infrequently; occurs in the dry N.W. Pacific to 980 ft (300 m). Spends most of the year underground, surfacing only during the first heavy rains of the rainy season to breed. Explosive breeder; many individuals emerge simultaneously in search of mates. Deposits eggs in temporary bodies of water. Tadpoles hatch within a few days and often form large schools of several thousand individuals. Male advertisement call is a bellowing *uwooooa*, likened to the sound of a person vomiting, hence Spanish common name *sapo borracho* (drunken toad). Calling male resembles a floating balloon when its internal vocal sacs are fully inflated.

**Family BUFONIDAE (Toads).** Members of this large cosmopolitan family occur on every continent except Antarctica. There are 18 species in Costa Rica. Fourteen of these are "true" toads, characterized by a chubby body, warty skin, and a poison (parotoid) gland behind each eye. They move with a hopping gait. All are nocturnal and terrestrial. Breeding takes place in pools and ponds, where they deposit their eggs in a long string. Four additional toad species occur in Costa Rica, all members of the genus *Atelopus*. Somewhat froglike in appearance, with slender frames, long legs, and smooth skin, they are diurnal and dwell in streams.

## Variable Harlequin Toad (*Atelopus varius*)

Male 1.5 in (3.9 cm); female 1.9 in (4.8 cm). Distinct sexual size-dimorphism. Smooth skin; pointed snout; angular body; long limbs; padlike hands and feet. Extremely variable in coloration and pattern: yellow, yellow-orange, or lime green, overlaid with black lines, spots, or blotches. In some populations, coloration changes with age. Formerly common at many sites at middle and high elevations, to 6,560 ft (2,000 m). Dramatic declines led to the disappearance of all but one population, and the species now teeters precariously on the brink of extinction. Terrestrial and diurnal. Lives on large rocks in the spray zone of high-gradient mountain streams. Sleeps on low vegetation or hidden in rock crevices. Bright coloration warns of powerful skin toxins.

## Litter Toad (*Bufo haematiticus*)

Male 2.4 in (6.2 cm); female 3 in (8.0 cm). Very large poison gland (parotoid gland) behind each eye; smooth skin; lacks bony crests on head that are typical of many toads. Light brown to purplish gray in coloration with dark brown lateral and ventral surfaces, creating the impression of a dead leaf. Red and orange blotches usually present on the hidden surfaces of the limbs. Young typically have white spots on the belly that disappear with age. Occurs in relatively undisturbed, humid forests on Caribbean slope and in S. Pacific, to 4,270 ft (1,300 m). Common but difficult to spot; found in leaf litter. Breeds between end of the dry season and middle of the rainy season, in rocky pools of streams and rivers. Young and adults retreat into surrounding forest when not breeding. *Illustration not to scale.*

## Giant Toad (*Bufo marinus*)

Other common name is Cane Toad. Can be enormous; largest females reach 9 in (23 cm) and weigh over 3.3 lb (1.5 kg). Male considerably smaller. Very large parotoid glands; warty skin; conspicuous bony crests on head. Back is gray, brown, or olive with indistinct blotches; belly dirty-white with irregular brown blotches. Most abundant around human settlements in hot lowlands, but found in a wide variety of habitats, to 5,250 ft (1,600 m); avoids closed-canopy forests. Nocturnal and terrestrial. Hides under cover during day; forages at night for invertebrates and small vertebrates. Skin glands contain powerful toxins that can debilitate or kill predators the size of an adult human. Breeds year-round in shallow margins of pools; egg masses contain 2,000 to 25,000 eggs (eggs and tadpoles are also toxic). Calling males produce a low-pitched trill. *Illustration not to scale.*

**Family CENTROLENIDAE (Glass Frogs).** A family of small, translucent, arboreal frogs. There are 13 species in Costa Rica. Flattened body; large forward-facing eyes (on tree frogs, the eyes are directed toward the side); translucent-green coloration, generally marked with white, cream, yellowish, or dark spots. Ventral surfaces are transparent and internal organs are often visible. Inhabit vegetation along banks of fast-flowing mountain streams. Most active on rainy nights. Females deposit eggs on either the upper or lower surfaces of leaves that overhang water; males call from leaves. Tadpoles generally hatch during heavy rains and drop into the stream below, where they complete their development.

## Emerald Glass Frog (*Centrolene prosoblepon*)

1.2 in (3.1 cm). Relatively robust. Emerald green, marked with small dark spots; white belly (a white sheath covers the intestines, making the belly less transparent than in most other glass frogs); bones bluish green. Adult male has a projecting spur on upper arm. Common (generally more abundant at higher elevations); occurs on Caribbean and Pacific slopes to 6,230 ft (1,900 m). Absent from dry N.W. Pacific. Usually seen on rainy or misty nights, perched on low vegetation along stream banks. Active year-round but breeding activity peaks during rainy season. Call a sharp series (usually three notes) produced at dusk and in early evening: *dik-dik-dik*. Deposits dark-colored eggs on upper surfaces of leaves, mossy rocks, and branches, up to 10 ft (3 m) above water. Males highly territorial; use spurs on arms to wrestle intruding males off leaves.

## Cascade Glass Frog (*Cochranella albomaculata*)

1.3 in (3.2 cm). Long limbs. Dark green to bluish green, covered with many cream to yellow bold spots; white lip stripe; dark-green bones. A white sheath covers the heart and many of the organs, but leaves much of the digestive tract visible through the ventral skin. Relatively common; occurs on Caribbean and Pacific slopes to 4,920 ft (1,500 m). Found in humid forests, on dense vegetation along streams. Encountered year-round but most active on rainy nights. Especially common in the spray zone of waterfalls. Call consists of a single, explosive *dik*, repeated at intervals of up to several minutes. Eggs are dark brown; lays clutch (generally horseshoe-shaped) on upper surface of leaves or on vertical rocks, just above the surface of the stream below.

## Fleischmann's Glass Frog (*Hyalinobatrachium fleischmanni*)

1.3 in (3.2 cm). Small and compact. Lime green with small yellowish-green spots. Ventral skin transparent, but heart, liver, and digestive tract are covered by an opaque white membrane. White bones. Common and widespread; occurs on Caribbean and Pacific slopes from 330 to 5,580 ft (100 to 1,700 m). Absent in dry N.W. Pacific. Found in moist and wet forests; tolerates some changes to its habitat and may persist in plantations and near human settlements. Reproductive activity peaks during rainy season. Male advertisement call is a single, whistled *wheet*; male calls from underside of leaves, several yards above water surface. Circular clutch of 20 to 50 small eggs (cream to greenish) deposited underneath leaf. Eggs are left unattended during the day but at night the male protects them from predators and keeps them moist.

# Variations on a Theme

Valerio's Glass Frog (*Hyalinobatrachium valerioi*) and eggs.

The female Valerio's Glass Frog lays a sticky mass of about 35 eggs on the underside of a leaf. In turn the male is responsible for guarding the clutch around the clock. At night he clambers over the eggs and urinates on them to keep them moist. During the day, he stands vigil, keeping one hand on the clutch as he faces the eggs. His spotted body resembles the egg mass next to him and perhaps functions as a decoy for predatory wasps and parasitic flies. Should either wasp or fly approach the male too closely, predator becomes prey. When the eggs hatch, the tadpoles fall into the water below, where they develop into froglets. (Not described within the species accounts.)

Rain frogs hatch as tiny froglets rather than tadpoles.

Rain frogs are a highly successful group of over 600 species in several genera. They employ a different developmental strategy. The female lays eggs in a suitable moist area, and one of the parents guards the eggs. The froglets develop inside the egg, skipping the free-swimming tadpole stage entirely. This strategy allows rain frogs to colonize areas far from the bodies of water that most other amphibians depend on for reproduction. This is undoubtedly a factor in their remarkable diversity.

**Family DENDROBATIDAE (Poison Dart Frogs).** Small, diurnal frogs that inhabit leaf litter. Noted for spectacular coloration and elaborate forms of parental care. The brightly colored species generally advertise powerful skin toxins, derived from a diet of alkaloid-rich invertebrates. Most species with cryptic coloration lack such toxins. None of the eight Costa Rican poison dart frogs pose a threat to humans. Some of these frogs are among the most easily observed amphibians in Costa Rica.

## Green-and-Black Poison Dart Frog *(Dendrobates auratus)*

1.7 in (4.2 cm). Medium size. Distinctive color pattern of black spots and turquoise-green to yellow-green spots. Generally common to abundant; occurs on Caribbean slope and in S. Pacific to 1,970 ft (600 m). Favors wet lowland forests but tolerates dense secondary growth and shaded plantations. Often hidden in leaf litter; most active in early morning and late afternoon.

Male generally nonterritorial; his call, a fairly high-pitched, insectlike buzz (*cheez-cheez-cheez*), serves to attract females. A clutch of 3 to 13 eggs is deposited in moist leaf litter. After fertilizing the clutch, male guards eggs until they hatch; he then carries larvae, one at a time, to an elevated water-filled basin. Tadpoles feed on algae and detritus, developing into dull-colored replicas of their parents within 7 to 15 weeks.

## Strawberry Poison Dart Frog *(Oophaga pumilio)*

1 in (2.4 cm). Small, relatively stocky; short limbs; smooth skin. Orange to scarlet, often marked with diminutive dark streaks or spots, but pattern variable. Hind limbs and forearms patterned with bright-blue or purplish-blue spots on a black background. The Granular Poison Dart Frog (*Oophaga granulifera*, formerly *Dendrobates granuliferus*) is similar in appearance but occurs only in the S. Pacific. Abundant and widespread; occurs on the Caribbean slope to 2,620 ft (800 m). Found in almost any habitat containing leaf litter and some tree cover. Conspicuous. Male call an insectlike *buzz-buzz-buzz*, repeated seemingly incessantly from a low perch. Eggs deposited in moist leaf litter. Male guards and tends eggs until they hatch. Female transports tadpoles to water-filled leaf bromeliad or other suitable plant and returns over the course of 9 to 10 weeks to bring food (unfertilized, nutritional eggs) to her developing offspring. (Formerly *Dendrobates pumilio*.)

### Striped Poison Dart Frog (*Phyllobates lugubris*)

1 in (2.4 cm). Small. Black with a pair of conspicuous yellow or pale-orange stripes; yellowish-green mottling on limbs. (The Golfodulcean Poison Dart Frog, *Phyllobates vittatus*, is similar in appearance but occurs only in the Golfo Dulce region.) Not uncommon but seen infrequently; occurs on Caribbean slope to 1,970 ft (600 m). Elusive, rarely venturing far from its hiding place in shaded sections of forest floor. Most active at dawn and dusk and on overcast, humid days. Male usually establishes his territory near fallen log or dense thicket of vegetation; his call is a low, buzzy trill that continues for several seconds. Eggs are deposited in leaf litter. Male hydrates eggs; after they hatch, he transports the tadpoles on his back to a nearby forest stream, where they complete their development.

### Common Rocket Frog (*Silverstoneia flotator*)

0.7 in (1.8 cm). Tiny. Grayish-brown back; chocolate-brown to black lateral surfaces; a distinctive pale lateral line is invariably present. In Pacific slope populations, pale lateral line extends to back of eye; on Caribbean slope populations, pale line ends before eye, not extending much beyond the groin region. Common; occurs locally on Caribbean and Pacific slopes to 2,790 ft (850 m). Found in relatively undisturbed humid forests. Usually seen moving about actively in leaf litter on steep banks of streams. Males very vocal; produce a slow, trilling whistle at dawn and dusk and after daytime rains. Deposits eggs in leaf litter. Several tadpoles are transported simultaneously on male's back to nearby stream to complete development; they float near surface of water and filter food particles with a large, funnel-shaped mouth. (Formerly *Colostethus flotator*.)

Pacific race

**Family HYLIDAE (Tree Frogs).** A large, cosmopolitan family of predominantly arboreal frogs; 42 species occur in Costa Rica. Tree frogs have conspicuously enlarged adhesive disks on fingers and toes; hands and/or feet are often webbed. They range in size from small to very large (females are generally considerably larger than males). Highly variable in appearance and biology, though all species are strictly nocturnal. Inhabit treetops, where they remain hidden from sight; generally only observed during breeding season, when they descend close to ground level. Breeding males call on rainy or humid nights, often in the vicinity of a body of water.

## Red-eyed Tree Frog (*Agalychnis callidryas*)

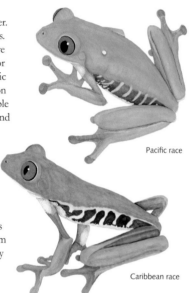

Pacific race

Caribbean race

Female 2.8 in (7.1 cm); male considerably smaller. Moderately large with exceedingly slender limbs. Red irises with vertical pupils; hands and feet are bright orange; pale vertical bars (cream, yellow, or white) on flanks. On individuals from many Pacific slope populations, flanks are purplish brown; on Caribbean slope frogs, flanks are blue to bluish purple (and upper arms and legs show blue). Abundant and widespread; occurs on Pacific and Caribbean slopes to 3,280 ft (1,000 m); absent from dry Nicoya Peninsula and rest of N.W. Pacific. Generally stays concealed in treetops, but when rains commence frogs move back and forth between canopy and breeding ponds. Males emit a short single or double *chuck*. Female lays eggs on vegetation overhanging water; parents sometimes fold leaf over eggs to protect them from sunlight. Tadpoles characteristically swim in nearly vertical position at surface of water, using only threadlike tail tip for locomotion.

## Parachuting Red-eyed Tree Frog (*Agalychnis saltator*)

Female 2.6 in (6.6 cm); male considerably smaller. Medium size. Red irises with vertical pupils; uniform blue or purplish-blue flanks. Leaf-green dorsal coloration changes to tan or brown at night. Relatively uncommon; occurs in isolated populations on Caribbean slope to 4,270 ft (1,300 m). Found in treetops of undisturbed forest; descends only after periods of heavy rains to breed in temporary pools and swamps; often forms large breeding aggregations. Breeding takes place in early morning hours; joined pairs deposit egg clutches on moss-covered vines and vegetation overhanging water. Known to "parachute" from considerable heights, using extensive webbing between spread fingers and toes to direct trajectory and slow descent.

# Explosive Breeding

Gliding Leaf Frogs (*Agalychnis spurrelli*) on the Osa Peninsula.

The Gliding Leaf Frog spends most of its life in the forest canopy, far from human observation. But on a given night during the rainy season—from May to October—a remarkable thing happens. En masse the frogs descend to temporary ponds to breed.

Launching from high in the canopy, they use the extensive webbing on their hands and feet to make controlled glides and turns, generally aiming for a large springy leaf on which to land. Up to several hundreds of frogs may congregate at just one pond. Mating takes place on a single night, during the course of which males emit a low-pitched moan to attract females. Because males significantly outnumber females, they must frequently fight for access to mates. On the day following the breeding frenzy, all frogs return to the canopy. Females deposit clumps of jelly containing roughly 15 to 60 eggs on the upper surface of leaves. Their tadpoles develop within the eggs for six days or so, then wiggle out and drop into the pond below. After two or three months, froglets exit the water and climb up to the canopy for the first time. (Not described within the species accounts.)

**Blue-sided Tree Frog** (*Agalychnis annae*)

Female 3.3 in (8.4 cm); male considerably smaller. Large but relatively slender; hands and feet are large and extensively webbed. Vertical pupils, yellowish-orange irises; leaf-green back; blue-and-pink coloration on the sides and extremities. Endemic to Costa Rica. Once relatively common, it has all but disappeared from much of its range, including in protected areas. Used to occur in the Tilarán, Central, and northern Talamanca mountain ranges from 2,560 to 5,410 ft (780 to 1,650 m), but now only known to exist in parks, gardens, and coffee plantations of the Central Valley. Breeds predominantly during the rainy season; lays its eggs on the upper surfaces of large leaves that overhang ponds. Male calls from a relatively high perch near breeding pool; produces a single *wooorp* note, repeated at intervals of several minutes.

**Lemur Leaf Frog** (*Hylomantis lemur*)

Female 2.1 in (5.3 cm); male smaller. Moderate size; very slender. Vertical pupils, silvery-white to bronze irises. Pale to lime green during the day, but changes to tan, orange, reddish brown, or chestnut at night. Flanks, upper arms, and thighs yellow to orange. Hands and feet unwebbed. Now very rare; formerly occurred on the Caribbean slope, with isolated populations in the N.W. Pacific, between 1,440 and 5,250 ft (440 and 1,600 m). Has suffered dramatic declines over the last two decades and only a few populations remain. Arboreal; generally walks or climbs on low vegetation in a studied, hand-over-hand fashion. Breeds during rainy season; male produces a short clicking call to advertise himself to females. Eggs deposited on upper surface of leaves overhanging shallow pond or swamp. (Formerly *Phyllomedusa lemur*.)

### Rufous-eyed Stream Frog (*Duellmanohyla rufioculis*)

Female 1.6 in (4.0 cm); male smaller. Medium size. Red iris with horizontal pupils. Extremely variable in coloration. Dorsum brown to mottled mossy-green (never bright green as in most red-eyed frogs). Distinct white stripe (along upper lip, flanks, and groin) expands below each eye to form a prominent white mark. Endemic to Costa Rica. Uncommon; occurs on Caribbean and Pacific slopes from 2,130 to 5,250 ft (650 to 1,600 m). Favors humid forests. Most frequently seen at night perched on vegetation beside streams. Males produce a soft, rasping call. Breeding biology unknown but tadpoles have an enlarged oral disk that enables them to cling to rocks in fast-flowing streams.

### Scarlet-webbed Tree Frog (*Hypsiboas rufitelus*)

Female 2.2 in (5.5 cm); male smaller. Medium size. Bluish green to dark green, generally with bold white spots and scattered dark specks; bright-red webbing between fingers and toes. Large eyes, horizontal pupils, silver to bronze irises. Green bones. Common at some sites but rarely seen; occurs on Caribbean slope to 2,130 ft (650 m). Found in forested lowland swamps; favors shallow, muddy wetlands with dense vegetation over open-water ponds, the preferred habitat of many other Costa Rican frogs. Only emerges from arboreal retreats on rainy wet-season nights. Deposits eggs in a floating raft, usually in shallow sections of breeding pools. Males produce a series of high-pitched clucks from perch in dense vegetation near water. (Formerly *Hyla rufitela*.)

### Gladiator Tree Frog (*Hypsiboas rosenbergi*)

Female may exceed 3.1 in (8.0 cm); male considerably smaller. Large. Diffuse, dark vertical bars on flanks; back and head mottled-cream, tan, brown, or reddish brown; a diagnostic black line runs down the middle of the back. Young frogs are pale green with small black spots. Relatively common; occurs in isolated populations on S. Pacific to 2,950 ft (900 m). Often found in flooded pastures and agricultural areas. Breeds year-round but breeding intensifies at end of dry spells. Male builds bowl-shaped depression at edge of stream or pond or uses existing water-filled hole. Eggs are deposited in a floating film on surface of water. Advertisement call (*tonk-tonk*) resembles the hammering of a stick on hollow log. Males aggressively defend their brood against trespassers. A sharp bony projection at the base of each thumb (most visible in males) is used to injure, sometimes kill, rival frogs. (Formerly *Hyla rosenbergi*.)

### Hourglass Tree Frog (*Dendropsophus ebraccatus*)

Female 1.4 in (3.5 cm); male smaller. Small. Tan to yellow with bold, dark-brown blotches. Dark dorsal markings usually form an hourglass-like pattern, though some individuals lack dark markings altogether. Common; occurs on Caribbean slope and, locally, in S. Pacific, to 5,250 ft (1,600 m). Favors relatively open areas in humid forests, parks, gardens, and agricultural areas. Breeds in shallow ponds at onset of rainy season. Male's advertisement call is a loud, insectlike *creek-eek-eek-eek*; call attracts females and also serves to ward off rival males. Eggs deposited on top surface of leaves overhanging ponds; tadpoles drop into the water below to complete their development. (Formerly *Hyla ebraccata*.)

### Small-headed Tree Frog (*Dendropsophus microcephalus*)

Female 1.2 in (3.1 cm); male smaller. Small. Yellow to tan overall, often with irregular dark lines and blotches on the back; a thin, dark-brown line (with narrow white border above) runs along the length of the body. The San Carlos Tree Frog (*Dendropsophus phlebodes*, formerly *Hyla phlebodes*) is similar in appearance but occurs only on the Caribbean slope. Abundant and widespread; occurs on Pacific slope to 3,610 ft (1,100 m). Found near flooded pastures, cattle ponds, and marshy areas. After rains, males call from emergent vegetation at the edge of breeding ponds. Its loud, insectlike advertisement call (*creek-eek-eek-eek*) is interspersed with clicking territorial call. Eggs are deposited in small clumps near the water surface, often attached to floating or emergent vegetation. (Formerly *Hyla microcephala*.)

### Meadow Tree Frog (*Isthmohyla pseudopuma*)

Female 2.0 in (5.2 cm); male slightly smaller. Medium size. Yellow tan to brown; dark specks on back fade during the day. Uniform cream, tan, or brown thighs; groin dark brown or blue with yellow spots. Breeding males tan with distinct dark-brown lateral stripe. Relatively common; occurs on Tilarán, Central, and Talamanca mountain ranges, between 3,610 and 7,550 ft (1,100 and 2,300 m). An explosive breeder after first heavy rains of rainy season; breeds in bodies of water as small as cattle hoof-print. Small egg clutches are attached to emergent vegetation in temporary bodies of water. Tadpole development is very rapid, an adaptation to life in small ephemeral ponds. Tadpoles become cannibalistic if supply of plant food is insufficient. (Formerly *Hyla pseudopuma*.)

breeding male

# Over Here!

An Hourglass Tree Frog (*Dendropsophus ebraccatus*) calls from a palm leaf.

On a warm, wet night in Costa Rica, visitors to swamps and ponds in lowland forests are likely to be greeted by a chorus of calling frogs. Frog calls, used to simultaneously attract mates and establish territoriality, can convey very complex information. Subtle differences between the call of one male and another, for example, allow a female to determine which would be the more suitable mate; louder, more complex calls often indicate a higher-quality male. But there are risks to the caller as well. Fringe-lipped Bats and other bat species can distinguish between the calls of edible frogs and poisonous frogs, and home in on the edible varieties. Frogs, in turn, have found ways to thwart such attacks. By keeping their calls short—and delivering them at irregular intervals—they make it more difficult for predators to pinpoint their location. Another strategy is to form a frog chorus. The cacophony confuses predators, and the source of a given individual's call gets lost in all the noise.

## Olive Tree Frog (*Scinax elaeochrous*)

Female 1.6 in (4.0 cm); male slightly smaller. Flat-bodied with long snout. Color and pattern on back is variable: some individuals are uniform gray, tan, or olive; others spotted or striped. During breeding season, adult male is bright yellow. Undersurfaces white or cream; dark-green bones are usually visible. Stauffer's Tree Frog (*Scinax staufferi*) is similar in appearance but it is smaller and has white bones; it replaces the Olive Tree Frog in the dry N.W. Pacific. Very common; occurs on Caribbean and Pacific slopes to 3,940 ft (1,200 m). Favors humid regions (absent from dry N.W. Pacific). Tolerant of changes to habitat; often found in disturbed areas and on walls of buildings. Active at night, during all but the driest periods of the year. Males call from vegetation near pond edges; advertisement call is a short, soft *waack*. Deposits eggs in a large mass that is often attached to floating water plants. Tadpoles live within vegetation in shallow margins of pond.

## Drab Tree Frog (*Smilisca sordida*)

Female 2.5 in (6.4 cm); male considerably smaller. Medium size. Dull coloration and extreme variation within populations make identification a challenge. Generally tan, gray, or reddish brown. Often uniformly colored, but sometimes has bold dark blotches or, less frequently, broad transverse bands across the back. Common; occurs on Caribbean and Pacific slopes (including Central Valley) to 4,920 ft (1,500 m). Absent from dry N.W. Pacific. Breeds during dry season in low-gradient streams and rivers. Males call at dusk from river banks or from boulders in rivers; they produce a short rattling advertisement call: *wrink*. In the breeding season, sleeping individuals can be found during the day clinging to river rocks.

## Masked Tree Frog (*Smilisca phaeota*)

Female 3.1 in (7.8 cm); male considerably smaller. Large. Has a conspicuous silvery-white stripe on the upper lip and a dark-brown mask. On most frogs, an irregular green mark separates the lip stripe and mask. Upper surfaces usually tan during the day, but can change to green at night. Common and widespread; occurs to 3,610 ft (1,100 m) in all but the driest regions of Nicoya Peninsula and the rest of the N.W. Pacific. Tolerates habitat alteration; prefers forest edges, roadside ditches, and secondary vegetation. When it rains, male calls at dusk from temporary pools, drainage ditches, tire ruts, etc. While floating, male produces a very loud, harsh advertisement call: *wrauk*. Female deposits as many as 2,000 eggs in small floating rafts. Eggs hatch after one or two days; tadpoles, capable of surviving up to 24 hours out of water if pool temporarily dries up, metamorph into frogs after about one month.

### Crowned Tree Frog (*Anotheca spinosa*)

Female 3.2 in (8.0 cm); male considerably smaller. Large. Series of sharp, bony spines on back of head is diagnostic (small juveniles lack spines). Cream or white encircles blotches and spots of light gray, brown, and dark brown. Uncommon; occurs in isolated populations on Caribbean slope from 330 to 6,560 ft (100 to 2,000 m). Found in undisturbed, humid forests. Highly arboreal and therefore rarely observed; possibly more widespread than commonly assumed. Breeds in water-filled tree cavities. Female produces unfertilized eggs to feed to tadpoles. Hard bony head and spines provide protection from predators when frog retreats into tree hole.

### Milk Frog (*Trachycephalus venulosus*)

Female 4.5 in (11.4 cm); male considerably smaller. Very large. Thick glandular skin. Beautiful gold-and-black iris. Extremely variable; upper surfaces usually have large dark spots on background of gray, yellow, reddish, or brown. Transverse bands on limbs. Common; occurs on Pacific slope to 980 ft (300 m). Found in a variety of open habitats, often in close proximity to humans. Arboreal and nocturnal. Survives dry months hidden in humid tree holes, leaf axils of bromeliads, and spaces under the bark of standing trees. Explosive breeding (in temporary rain pools) triggered by first rains of the rainy season. Mating and egg-laying restricted to a few rainy nights every year. Skin glands produce a milky, white secretion that is extremely sticky and toxic. Can cause pain— even temporary blindness—if rubbed in eyes. (Formerly *Phrynohyas venulosa*.)

**Family HEMIPHRACTIDAE (Marsupial Frogs).** An essentially South American family with only a single species that reaches Costa Rica. These frogs display direct development (the tadpoles develop entirely within the egg without a free-swimming larval stage) and they carry the developing eggs on their back until they hatch. The taxonomic status of this group is uncertain; marsupial frogs were long considered a subfamily within the family Hylidae, although some biologists believe them to be more closely related to the family Leptodactylidae.

### Horned Marsupial Frog (*Gastrotheca cornuta*)

3.2 in (8.1 cm). Large. Unique in having flaplike projections over the eyes and transverse ridges across the head and back; tan or brown (at night coloration becomes darker). Only known from a handful of sightings; occurs locally on S. Caribbean slope, to 3,280 ft (1,000 m). Arboreal. Female carries up to 9 fertilized eggs in a skin pouch on her back until fully developed froglets hatch; this species does not require bodies of water for reproduction and is thus able to remain hidden in the canopy even when breeding.

**Family LEPTODACTYLIDAE (Rain Frogs and relatives).** A large family of diverse frogs, many of which vary greatly in appearance and are difficult to identify. The 47 Costa Rican species in this family include both some of the smallest and largest frogs in the country. Females of the genera *Craugastor* and *Eleutherodactylus* deposit a small clutch of eggs in a moist site; the larvae transform into tiny froglets within the egg, skipping a free-swimming tadpole stage. Frogs in the genera *Engystomops* and *Leptodactylus*, however, deposit their eggs in floating foam nests that protect the eggs from predation and dehydration. Upon hatching, these tadpoles drop into the pond below and complete their development in water. Drastic taxonomic changes are afoot and this family will likely be split into several smaller families in the future. A recent change already resulted in a split of the genus *Eleutherodactylus*, and many of its Costa Rican species are now placed in the genus *Craugastor*.

## Common Rain Frog (*Craugastor fitzingeri*)

Female 2.1 in (5.3 cm); male considerably smaller. Medium size. Color pattern extremely variable: gray, beige, or brown, often marked with either a yellow or pale-orange stripe down the back or with a dark spot (with hourglass or W-shape). Very difficult to identify in the field using coloration as key. Abundant and widespread; occurs on Caribbean and Pacific slopes to 4,920 ft (1,500 m). Favors grassy clearings and banks of streams and rivers. Male advertisement call resembles sound of someone rapidly clicking two pebbles together; frequently heard at dusk and during strong daytime rains. Often perches at night, exposed on low vegetation. The Fringe-lipped Bat (*Trachops cirrhosus*) is known to locate this frog by its call and eat it. *Illustration not to scale.*

## Bransford's Litter Frog (*Craugastor bransfordii*)

1 in (2.6 cm). Tiny. Dorsum tan, gray, cream, or brown with a highly variable pattern of marks. Skin coloration, pattern, and texture vary greatly and are not good field characteristics; most easily identified by complete absence of disks and webbing on fingers and toes. Locally abundant on Caribbean slope to 2,890 ft (880 m). Found in leaf litter of humid forests; most common at the end of the dry season when leaf litter is thickest. A successful colonizer of disturbed areas; prefers forest edges, trails, and clearings rather than forest interior. Male advertisement call is a soft *pew*.

## Atlantic Broad-headed Rain Frog (*Craugastor megacephalus*)

Female 2.8 in (7.0 cm); male much smaller. Medium to large. Has a very broad head (whose width approaches or exceeds half of the animal's length). Bony ridges on snout, between the eyes, and on back. Reddish, gray, tan, or dark brown, often with dark countershading on ridges. Relatively common (young seen more frequently than adults). Occurs on Caribbean slope to 3,940 ft (1,200 m). Inhabits floor of humid and moist forests, usually concealed in leaf litter. Sit-and-wait predators; at night adults ambush passing prey, dashing out from burrows to capture spiders and other relatively large invertebrate prey. *Illustration not to scale.*

**Common Tink Frog** (*Eleutherodactylus diastema*)

0.9 in (2.4 cm). Tiny. Highly variable in coloration and pattern—and individuals are capable of considerable color change. During the day, gray to brown, with variable markings; at night, uniformly pink or tan. Abundant and widespread; occurs on Caribbean and Pacific slopes to 5,310 ft (1,620 m); absent from dry N.W. Pacific. Generally walks or runs rather than jumps. Strictly arboreal; female deposits eggs in treetop bromeliad or other moist spot. Young exit eggs as miniature copies of parents, skipping free-swimming tadpole stage. Difficult to spot and to identify but often heard. Its advertisement call, a loud clear *dink*, is one of the most common night sounds. Male usually calls while perched upside down on the underside of a leaf.

**Pygmy Rain Frog** (*Eleutherodactylus ridens*)

Female 1 in (2.5 cm); male smaller. One to several pointed tubercles on top of each eye. Raised nostrils. Generally has pale coloration: gray, tan, yellowish, or pink, with a dark mark behind the eye (covering the top of the ear drum). Red on thighs, calves, and feet not visible when legs are folded. Very common and widespread; occurs on Caribbean and Pacific slopes to 5,250 ft (1,600 m); favors humid environments, absent from dry N.W. Pacific. Somewhat tolerant of changes to habit; found in both pristine forests and plantations and gardens. Hides during the day in leaf litter, moss mats, and arboreal bromeliads, but emerges at dusk. Usually seen at night, perched on low vegetation. Male produces a chuckling, harsh trill.

**Tungara Frog** (*Engystomops pustulosus*)

1.4 in (3.5 cm). Small. Toadlike with wartish pustules on upper surface of head, body, and limbs. Distinct parotoid glands are present. Coloration highly variable: generally shows shades of brown and gray with irregular dark markings on the back and light blotches on the limbs. Usually has a distinct pale line down the middle of the throat and belly. Occurs in the N.W. Pacific and S. Pacific (absent from Central Pacific) to 4,920 ft (1,500 m). Often found in close proximity to human settlements. Heard on rainy nights calling from shallow basins that contain water and some emergent vegetation. Male advertisement call is an explosive *mew*. As do species of the genus *Leptodactylus*, this frog produces a foam nest in which to lay eggs; tadpoles develop in the water below the nest. Retreats underground during the dry season, and is rarely seen then. (Formerly *Physalaemus pustulosus*.)

**Black-backed Frog** (*Leptodactylus melanonotus*)

2.2 in (5.5 cm). Medium size, robust. Males identified immediately by a pair of black spines at the base of each thumb. Uniformly dark brown or with a pattern of dark blotches, bands, or stripes; usually has a dark triangular mark on top of the head; lips show alternating dark and light bars. Very common; occurs on Caribbean and Pacific slopes to 4,270 ft (1,300 m). Prefers open areas, where it inhabits bodies of water sometimes as small as a water-filled hoof print. On humid nights male calls from concealed location in vegetation or in burrow beneath rocks or logs; advertisement call is a soft, mechanical *tuc-tuc-tuc*. Tadpoles of this species often form large schools.

**Smoky Jungle Frog** (*Leptodactylus savagei*)

7.3 in (18.5 cm). Massive. Tan, reddish brown, brown, or purplish gray, either uniformly colored or with a pattern of squarish spots and bars that are faintly outlined with dark pigment. Male has a large black spine on each thumb and a pair of smaller spines on the chest; the male uses these spines in territorial battles with other males and to grip female when mating. Breeding male has greatly swollen arms. Common and widespread; occurs on Caribbean and Pacific slopes to 3,940 ft (1,200 m). Generally seen after dark, perched at edge of pond or swamp. At dusk and into early evening, male produces a loud, somewhat ominous, *wrrrooop* call as he sits in water-filled burrow. When picked up, emits a bone-chilling scream and secretes slimy skin secretions that can cause rashes and a burning sensation in humans. (Formerly *Leptodactylus pentadactylus*.) *Illustration not to scale.*

**Family MICROHYLIDAE (Narrow-mouthed Frogs).** All three species of narrow-mouthed frog that occur in Costa Rica spend much of their time underground. An egg-shaped body, smooth skin, small head, and powerful hind feet with enlarged, spadelike tubercles are all adaptations to a burrowing lifestyle. They surface in the rainy season to breed.

**Sheep Frog** (*Hypopachus variolosus*)

Female 2.1 in (5.3 cm); male considerably smaller. Small to medium size. Body egg-shaped, with a distinct transverse skin fold running across the top of a small, pointed head. Two prominent spadelike tubercles on each foot. Gray, olive, brown, or reddish brown, with scattered dark spots; a thin yellow or orange line runs down the back. Common but seen infrequently; occurs in dry N.W. Pacific and, locally, in the Central Valley, to 5,250 ft (1,600 m). Forages mostly on ants and termites. After heavy rains, emerges from its burrow at night to reproduce in temporary pools. Males produce a bleating, sheeplike advertisement call. Eggs, deposited on surface of water in a floating film, hatch rapidly; tadpoles also develop quickly (in about one month), an adaptation to life in small breeding pools that can dry up from one day to the next.

**Family RANIDAE (True Frogs).** Members of this family occur throughout most of the world. Characterized by a streamlined body and long, muscular hind legs; most species have extensive webbing between the toes for propulsion in water. These frogs are strong swimmers and excellent jumpers on land. All five Costa Rican ranids have a skin fold that runs from behind each eye to the groin; and webbed feet (but no webbing on hands).

## Vaillant's Frog (*Rana vaillanti*)

Male 3.7 in (9.4 cm); female 4.9 in (12.5 cm). Very large. Has powerful legs and large feet with extensive webbing. Tan to dark brown overall; head and front half of body often green. Common and widespread; occurs on Caribbean slope and in N.W. Pacific (including the Nicoya Peninsula) to 2,630 ft (800 m). Diurnal and nocturnal. Lives on the banks of ponds, streams, and rivers; when startled, it escapes into the water. An accomplished swimmer. Male advertisement call resembles sound made when one rubs a finger over an inflated balloon. Deposits eggs in water in large clumps that may contain up to several thousand eggs. Tadpoles are very large (to 3.1 in/8.0 cm) and conspicuous, with a dark head and boldly spotted tail.

## Green-eyed Frog (*Rana vibicaria*)

Male 2.9 in (7.3 cm); female 3.6 in (9.2 cm). Large. Green irises and bright-red thighs are diagnostic; bronze or brown on back. Young frogs with bright-green back; changes gradually into adult coloration. Less aquatic than some of its relatives, as indicated by relatively small amount of webbing on feet. Once common in the Tilarán, Central, and Talamanca mountain ranges, between 4,920 and 8,860 ft (1,500 and 2,700 m); currently known only from a single breeding population near Monteverde. Male lacks both vocal sac and vocal slits; produces a soft trilling advertisement call that does not travel far. Breeds during rainy season, in forest ponds and in slow-moving sections of rivers; eggs deposited in the water as a globular mass.

## Brilliant Forest Frog (*Rana warszewitschii*)

Male 2 in (5.2 cm); female 2.5 in (6.3 cm). Medium size. Streamlined body and a long, pointed snout. Feet extensively webbed. Back is bronze, speckled with bright-blue or green spots; sides usually darker brown, especially in young frogs. Underparts of limbs bright red. Back of thighs dark brown with two to four bold yellow spots (not visible when the legs are folded). Suddenly exposing these bright flash colors presumably confuses or deters potential predators. Common and widespread; occurs on Caribbean and Pacific slopes to 5,580 ft (1,700 m). Favors humid forests; absent from the Nicoya Peninsula and the rest of the dry N.W. Pacific. Least aquatic of the Costa Rican true frogs; usually found in leaf litter, although rarely far from the ponds or slow-moving streams in which it breeds year-round. Eggs laid in a clump and attached to undersides of rocks; tadpoles become very large, to 4.5 in (11.5 cm).

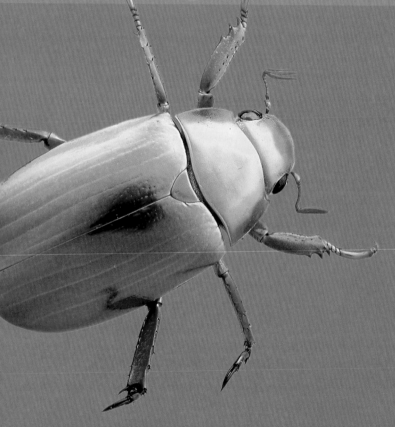

# arthropods

**The phylum Arthropoda** is a vast collection of animals whose characteristics include an external skeleton, segmented body, and jointed limbs. This group—as diverse as it is large—includes insects, spiders, scorpions, millipedes, and crabs and other crustaceans. In Costa Rica, the insects alone number from 200,000 to 250,000 species! This means, of course, that the species included in this chapter necessarily represent a mere fraction of the arthropods that occur in the country.

But even so, the reader will find in the following pages many of the most conspicuous, beautiful, or interesting insect species to be found in Costa Rica. Among them is the Blue Morpho (*Morpho helenor*), a striking butterfly that visitors are likely to see as it flits above rainforest trails. There are six species of morpho butterfly in Costa Rica, several of them similar in appearance. In the field, it is often very difficult to distinguish between the species within some genera—even for biologists—and in such cases a general description of the characteristics of the genus is given in lieu of a species description. Four genera of ant species are described; ants are readily seen, fascinating, and ecologically important—they are thought to make up a very large percentage of the insect biomass of Costa Rica. Leaf-cutter ants carrying snippets of vegetation are impossible to miss in lowland forests, and their story is one of several interesting examples of insect societies in action.

Other arthropods featured in this chapter include spiders, scorpions, and crabs, many of which are conspicuous and common. First-time visitors to the tropics may be surprised to see crabs inland, often a considerable distance from water. The Velvet Worm is also included—even though it is in the phylum Onychophora—because it is unique and because of its resemblance to millipedes.

During the day, it's easy to see arthropods in Costa Rica: they are multitudinous and they are everywhere. To view insects at night, however, try suspending a white sheet and setting up a light behind it; the light will attract a variety of insects—including moths, dobsonflies, and beetles—that will come to rest on the sheet. And pay close attention to animal dung—it is a rich food source for various beetles. When daubed on trees, either ripe fruit or a mixture of fruit, sugar, and beer will attract many species. Cryptic species often can be found by searching under large leaves, inside coiled leaves, and in leaf litter.

Jewel Scarab (*Chrysina chrysargyrea*). The amazing color of this beetle is produced not by pigment but rather structurally, via interference of light waves on microscopic structures inside the forewings.

# An Ancient Lineage

Helicopter Damselfly (*Megaloprepus coerulatus*).

Damselflies and dragonflies, in the order Odonata, have winged their way across the surface of this planet for some 400 million years. This ancient lineage of insects, along with one other group, the mayflies (Ephemeroptera), is equipped with permanently outstretched wings that cannot be folded and stored on their back. These insects are collectively labeled Paleoptera ("old-winged"). All other winged insects—the Neoptera ("new-winged")—are able to fold up their wings like an umbrella and store them safely when not in flight; they make up the vast majority of insect species in Costa Rica and worldwide.

The Helicopter Damselfly has the largest wingspan of all living odonates, with a maximum length of 7.5 in (19 cm). Another striking feature is its flight pattern—each wing beats independently of the other wings, with a helicopter- or windmill-like motion. This damselfly feeds on orb-weaving spiders, which apparently do not detect the slender, transparent form of the damselfly as it hovers in front of their web. In a single move, the damselfly flies in and cuts off the spider's abdomen, allowing the rest of the spider to fall into the web or onto the ground. (Not described within the species accounts.)

## Order Phasmatodea (Stick Insects)

### Guanacaste Stick Insect (*Calynda bicuspis*)

5 in (13 cm). Very distinctive body shape; green or brown in color, with highly variable patterning. Adults of both sexes are wingless. Very common in savanna habitats of dry N.W. Pacific. Found on low shrubs. During the day, remains motionless and is very easily overlooked. At night, searches for food and mates. Mating male uses hooks on his abdomen to cling to the lower abdomen of the female, and he sometimes remains attached for several days. Female produces eggs continuously, flipping each egg to a point on the ground below a food plant; she lays up to 12 eggs per night. Eggs—seedlike in appearance—are often picked up and carried by ponerine ants. Lifespan is 6 months to 1 year.

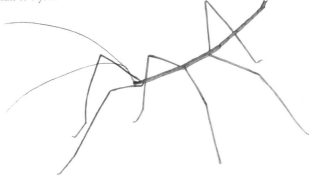

## Order Isoptera (Termites)

### Arboreal Termite (*Nasutitermes* sp.)

Worker 0.2 in (0.5 cm); queen substantially larger. Occurs in rainforests and at forest edges, at low elevations. Conspicuous, spherical, blackish nests—the size of a football or larger—are usually situated on a tree trunk or limb. Nest appears to consist of dark soil but is actually made from chewed wood-fiber that is glued together with fecal material. Each colony is composed of 3 castes: reproductives (one queen and king), sterile workers, and sterile soldiers. Workers and soldiers can be male or female. Worker termites are blind and navigate using chemical signals. They make covered trails to food sources, and if these trails are disturbed the workers quickly disappear and soldiers rush out to defend the nest. Soldiers have an enlarged, pointed head that discharges turpentine-like secretions; these chemicals are toxic to small invertebrate predators, and irritating to anteaters. Winged, reproductive adults swarm at the beginning of the dry season. *Illustration not to scale.*

soldier termites

## Leaf Praying Mantis (*Choeradodis* sp.)

3 in (7.5 cm). This amazing leaf mimic has a flat, leaflike thoracic shield and a rounded, leafy wing case. It lies in wait for small arthropods, using its raptorial forelimbs to catch and manipulate prey. Contrary to popular lore, female mantids seldom decapitate their mates during copulation; in the rare cases that this does occur, however, the headless male continues to mate with the female, inserting a sperm packet into the female's abdomen. She uses this sperm package to fertilize her eggs, laying several cases of 50 to 100 eggs over the next few weeks. She attaches each egg case to the underside of a leaf or branch.

## Giant Cockroach (*Blaberus giganteus*)

3.5 in (9 cm). This is one of the largest cockroaches in the New World tropics. Cream brown with a dark spot on the head and a dark band across the wings. Exceedingly common; occurs throughout the country in both wet and dry forests. Nocturnal, sometimes attracted to lights. During the day it rests in hollow trees, caves, or rotten logs. Omnivorous; eats bat guano, fruit, and soft plant material. Mating occurs with male and female facing away from each other. Females internally incubate about 40 eggs for 60 days. The tiny young emerge pure white but turn brown as the exoskeletons harden in air. Early stages are wingless.

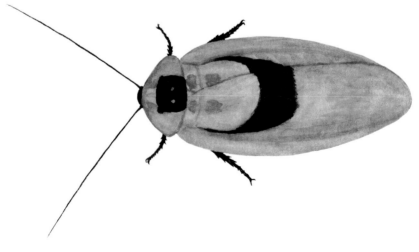

# In Disguise

Green Leaf Mantis (*Choeradodis rhombicollis*).

Many insects have evolved camouflage that allows them to blend in with their surroundings, but few do so as artfully as the leaf mantids. Some resemble living leaves, others take on the form of dead leaves, replete with curled-up, dry-looking wings and small holes that give the appearance of chew marks or partial decomposition.

Leaf mantids take up position on a branch covered in leaves of similar shape to their body. They remain motionless and sit flat against the branch, unlike other praying mantids, which retain a more upright posture. If disturbed, some species vibrate gently, mimicking the motion of a leaf stirring in the wind. Camouflage serves not only as a means of avoiding predators but also to conceal oneself from passing prey. Employing a sit-and-wait hunting strategy, mantids are catholic in diet and eat a variety of arthropods. Indeed, some larger mantid species have even been known to capture and consume hummingbirds and lizards.

### Walking-leaf Katydid (*Mimetica mortuifolia*)

2 in (5 cm). Forewings—with green or brown coloration, leaflike venation, and "bitten" edges—bear a striking resemblance to living or dead leaves. The hind wings are greatly reduced. Occurs at low and middle elevations. Found in forests, from understory to canopy. Camouflage makes it hard to spot. At night, feeds on leaves. Males make short buzzing calls to attract mates.

### Cone-headed Katydid (*Copiphora rhinoceros*)

1.5 in (4 cm). Spiny legs and cone-shaped spear on forehead are diagnostic. Occurs at low elevations in rainforests and second growth (from Central America to Brazil). Nocturnal; can be found by directing flashlight at understory vegetation. Unlike most katydids, this species is an active predator, feeding on snails, leaf katydids, and even small lizards; it also eats seeds. To attract females, male alternates loud, high-pitched chirps with full-body tremulations that cause vibrations in the surrounding plants. When a female approaches, he limits his display to silent shudders that are less easily detected by competing males—and predators such as bats. Female uses her long ovipositor to deposit eggs in deep pockets between leaves of bromeliads and palms.

### Sundown Cicada (*Fidicina* sp.)

1.6 in (4 cm). One of the largest cicadas in Costa Rica. Stout, fuzzy body; brownish patterning on the wings. Occurs on Caribbean and Pacific slopes at low and middle elevations. Found in both rainforest and dry forest. Although many cicadas are restricted to the forest canopy, this species clings to trunks of large trees within 10 ft (3 m) of the ground. Emits incessant, sirenlike call only in the late afternoon in rainforests (but in dry forests, it calls intermittently throughout the day). After mating, female lays oblong mass of white eggs on palm fronds or other leaves. On hatching, the nymphs drop to the ground and burrow into the soil to feed on juices of plant roots. Adults emerge once a year in tropical regions, often appearing en masse, leaving numerous cast skins on the ground. They live for a few weeks or months, feeding on plant juices.

### Thorn Bug (*Umbonia crassicornis*)

0.5 in (1.3 cm). This small bug bears a striking resemblance to a rose thorn. Adult is green or yellow with brown and red markings. Breeding female is dark green. Young has 3 horns. Widely dispersed; occurs in both wet and dry habitats. Prefers disturbed areas and second growth. It is common in gardens in San José, feeding on ornamental leguminous trees and shrubs, citrus trees, and various legumes. Female lays egg masses in branches or petioles of host plant. When the eggs hatch, the female makes a spiral slit around the bark of a branch, and the young move into this slit. Any nymphs (young thorn bugs) that stray beyond the slit are stopped by the mother. She continues to guard her young until they reach adulthood. Costa Rica has about 200 species of thorn bug (treehopper), several species of which exhibit maternal care. *Illustration not to scale.*

### Peanut-headed Bug (*Fulgora laternaria*)

4 in (10 cm). When this large, grayish-white insect sits motionless on a tree trunk, it blends in perfectly with the background. Only when grabbed or knocked off its perch, will it spread its wings fully, flashing two large, bright eyespots on the hind wings. Young are extremely similar to ants and also mimic antlike behavior. Occurs on Caribbean and Pacific slopes at low elevations; in the N.W. Pacific, it is almost always found on Stinking Toe trees (*Hymenaea courbaril*). Inhabits both dry and wet forests. When disturbed, it leaps, discharges a skunklike odor, and runs up its tree trunk. The *machaca*, as it is locally known, is a harmless bug yet the subject of many myths: it is falsely said to be luminescent, and, according to common lore, if a girl is stung by a *machaca*, she will die within 24 hours unless she sleeps with her boyfriend.

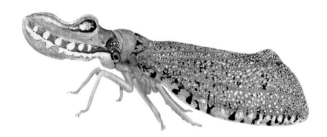

When citing forewing length, the number indicated is for a single forewing rather than for the combined length of the two forewings.

### Postman (*Heliconius erato*)

Forewing length: 1.4 in (3.5 cm). Long, black forewings show red stripes; slightly curved yellow stripes on hind wings distinguish this species from those with straight pale-yellow stripes. This is one of the most common (and easily recognized) heliconia butterflies in Costa Rica; occurs on Caribbean and Pacific slopes to 5,250 ft (1,600 m). Found in gardens, plantations, forest edges, and second growth, usually flying a few feet above the ground. Adult feeds on pollen from a variety of flowers. Female lays a single egg per day on the leaf tips of passion flowers (*Passiflora*). At night, groups of up to about 10 adults roost in low vegetation, often near water.

### Zebra Longwing (*Heliconius charitonia*)

Forewing length: 1.7 in (4.3 cm). Black narrow wings strikingly patterned with slender yellow stripes and a line of pale spots (other related species have only 3 stripes). Common; occurs on Caribbean and Pacific slopes to 3,940 ft (1,200 m). Found in disturbed areas and secondary forest, rarely in mature forest. Adult visits flowers of *Lantana*, *Hamelia*, and other plants for pollen and nectar. Female lays small clusters of eggs on passion flower vines. Groups of adults return to the same roost night after night.

### Hecale Longwing (*Heliconius hecale*)

Forewing length: 1.8 in (4.5 cm). Large, with relatively broad wings. Forewings rounded at tips. Color and pattern varies across range; some individuals lack yellow spotting or have bands of yellow and black on hind wings. Common and widespread; occurs on Caribbean and Pacific slopes to 5,580 ft (1,700 m). Found in a wide range of habitats. Probably migrates from Pacific to Caribbean slopes. Adult feeds on a variety of flowers and will defend food plants from other butterflies.

### Julia (*Dryas iulia*)

Forewing length: 1.7 in (4.3 cm). Male is easily recognized by elongated, bright-orange wings. Female is duller, with black markings on upper wings. Rapid, fluttery flight; travels about 10 ft (3 m) above the ground. Common; occurs on Caribbean and Pacific slopes to 4,920 ft (1,500 m). Prefers forest edges and open areas, but also found in the forest canopy. Female lays single egg on tendrils of passion flower vines or, sometimes, on adjacent plants. Adult feeds on nectar of a variety of flowers.

male

### Rusty-tipped Page (*Siproeta epaphus*)

Forewing length: 2 in (5 cm). Large. Wings with rusty-brown tips, a white stripe in the middle, and a coffee-black base. Occurs on Caribbean and Pacific slopes, from 1,310 to 4,920 ft (400 to 1,500 m). Found in evergreen forests, forest edges, and disturbed areas. Flutters in back-and-forth fashion up to about 6.5 ft (2 m) above the ground. Adults visit flowers of *Lantana, Impatiens, Croton*, and other species. Female lays clusters of eggs on Acanthaceae. Caterpillars are black and spiny and have prominent horns on head.

### Malachite (*Siproeta stelenes*)

Forewing length: 1.7 in (4.5 cm). Colorful. Upper side dark brown with translucent-green windows; underside grayish with pale-green windows. Common; occurs on Caribbean and Pacific slopes to 4,590 ft (1,400 m). Found in gardens, open second growth, and clearings in forest. May be encountered at rest with wings closed, or traveling in slow, floating flight. Adult visits flowers, dung, carrion, and rotting fruit. Female lays single, dark-green egg on plants. Caterpillars are black and spiny and have prominent horns on head.

### Pink-tipped Satyr (*Cithaerias pireta*)

Forewing length: 1.2 in (3 cm). Forewings transparent; hind wings are clear near body, washed with pink on edge, and marked with a bluish eyespot. This butterfly, sometimes called a glasswing or clearwing, is in the subfamily Satyrinae. Most other glasswings (subfamily Ithomiinae) have long, narrow wings and lack eyespots on the hind wing. Fairly common; occurs on Caribbean and Pacific slopes to 3,280 ft (1,000 m); on Pacific slope, occasionally found to 6,560 ft (2,000 m). For a period of several weeks, male tends to remain close to a specific forest light-gap. Seen alone or in small groups, usually in bouncy flight near the forest floor. Adult feeds on rotting fruit and fungi.

## Blue Morpho (*Morpho helenor*)

Forewing length: 2.8 in (7 cm). A large, iridescent-blue butterfly; underside of wings is chocolate brown, patterned with cream coloration and 7 eyespots. Caribbean slope populations are mostly blue above, with wings edged black. Pacific slope and Central Valley populations are mostly blackish brown above, with a narrow band of iridescent blue on the forewing. Common; occurs on Caribbean and Pacific slopes to 4,590 ft (1,400 m). Found in a wide variety of habitats. With characteristically slow, floppy flight, passes along trails and waterways. Adult feeds on rotting fruit and carrion. Female lays eggs singly on large leguminous trees such as *Pterocarpus* and *Lonchocarpus*. Development from egg to adult takes 2 to 3 months.

Caribbean slope individual

## Banded Owl Butterfly (*Caligo atreus*)

Forewing length: 3.1 in (8 cm). About 15 species of owl butterfly occur in Costa Rica, most of them with large eyespots on the underside of the wing. The Banded Owl Butterfly is distinguished by blue-purple color of the upper forewing and a cream band on the upper hind wing. Occurs on Caribbean and Pacific slopes to 4,270 ft (1,300 m); absent in dry N.W. Pacific. Found in rainforest and second growth. Most active at dawn or dusk, it rests on tree trunks during the day, displaying a single eyespot. Theories about the eyespots include the possibility that their resemblance to owl eyes may scare off potential predators. Adult feeds on rotting fruit, sap, dung, and carrion. Female lays eggs on heliconias, bananas, and cyclanths.

Owl Butterfly (*Caligo memnon*).

The Owl Butterfly is one of the largest butterflies in the world. As it wends its way through the forest understory with slow flaps of its large wings, its most salient feature is the blue and cream colors on the upper surface of the wings. On landing, it closes its wings, and now the brown underwing colors blend in relatively well with the branch on which it has come to rest. Many butterflies display brilliant colors in flight but show subdued colors when stationary—and most vulnerable to predators.

The wing eyespots themselves are perhaps an antipredator strategy, providing a target that lures predators away from the butterfly's body. Thus, with its body intact, the Owl Butterfly may be able to escape an attack bereft only of a piece of its wing. (Not described within the species accounts.)

## Cuatro Ventanas (*Rothschildia orizaba*)

Forewing length: 2.8 in (7 cm). This large silk moth has a transparent window on each of its four wings, hence the common name *cuatro ventanas* (four windows). The nontransparent surface of the wings is a warm-brown color with cream markings. Common; occurs on Caribbean and Pacific slopes to 4,920 ft (1,500 m), in all forest types. This species is sometimes seen fluttering around lights in the Monteverde region. Mates within a few hours of emergence, and never feeds during its brief life.

Breeding occurs from Feb. to April and from Sept. to Nov. Female lays eggs on several consecutive nights, placing them on *Salix, Prunus,* and other host plants. Caterpillars are bright green above, bluish green below, and with a white stripe on the side. The silken cocoon of the pupa can be harvested, but silk is principally made from Asian silk moths.

## Green Urania (*Urania fulgens*)

Forewing length: 2 in (5 cm). The Green Urania is a day-flying moth, not a butterfly. Noted for its metallic green-and-black coloration and long swallowtails. Most common in low-elevation rainforests of the Caribbean and Pacific slopes. Adult feeds on flower nectar, wet sand, and tree sap. Female lays large clutches of eggs on *Omphalea,* a canopy liana in the Euphorbia family. Breeding season begins in May and can continue until Dec. In some years, huge numbers of moths migrate through the Central Valley. Thousands may congregate before taking flight. The *Omphalea* vines appear to increase in toxicity when consumed by moths for several years; mass migrations presumably allow moths to seek out new host plants that are less toxic.

# In Plain View

Leaf-roller Moth (*Pseudatteria leopardina*).

Unlike most moths, which are nocturnal, the Leaf-roller Moth is active during the day, when it is most exposed to birds and other predators. Instead of concealing itself under a branch or folding away its bright colors under plain-colored upper wings, this moth quite openly puts itself on display. Its electric colors and busy pattern are its salvation, however, because they signal to potential predators that it is unpalatable. Most birds quickly learn to avoid poisonous or distasteful prey after having attempted to eat an unsavory individual or two, which means the birds have the capacity to store visual memories.

There are several examples of moths and butterflies that are brightly colored but in fact harmless. These species mimic a poisonous species in order to avoid being eaten, a strategy that works only if the distasteful individual is more numerous than the tasty copy. (Not described within the species accounts.)

### Dung Beetle (*Dichotomius carolinus*)

1.2 in (3 cm). Body entirely black, stocky, and convex. On dung beetles, the hind legs are located far back on the body. This species, the largest dung beetle in Costa Rica, ranges throughout the country at low and middle elevations. Nocturnal. Feeds on cow and horse droppings, but also eats other types of dung. Unlike other dung beetles, this species does not transport dung by forming dung balls and rolling them. Instead, it either carries small droppings to its burrow or, in the case of large droppings, digs a burrow right next to the dung. Within the burrow, adults feed on dung and also use it to form a brood mass, which contains one large egg. Development from egg to pupa takes 2 months; adults emerge in the rainy season. The Dung Beetle is sometimes seen near lights set up to attract moths.

### Hercules Beetle (*Dynastes hercules*)

6.7 in (17 cm). This, one of the largest beetles in the world, is the most striking of the beetles collectively known as rhinoceros beetles. Body is brown or yellowish. Male has 2 black horns that are as long as its body; the female lacks horns, but her body is larger than that of the male. Occurs on Caribbean and Pacific slopes at low and middle elevations. Nocturnal; adult is attracted to lights. Feeds on decaying plant material and fruit. Amazingly, the Hercules Beetle can carry at least 100 times its own body weight. Mates in the rainy season; male bobs head in threat display, makes a chirping sound, then uses his long horns to lift opponent and fling him to the ground. Larva eats decomposing material, growing for about 9 months before pupating in a case made of wood fiber; the mature larva is huge, white, with short, brown legs, and weighs up to 4.2 oz (120 g).

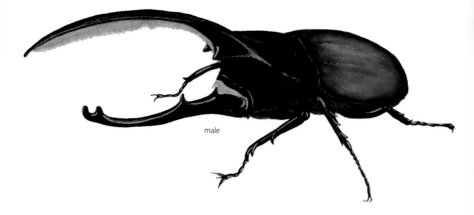

male

# Hercules Beetle

Male Hercules Beetle (*Dynastes hercules*).

One of the largest scarab beetles in the world, the Hercules Beetle is highly dependent on old-growth forests for its survival. The larval form of this beetle and related species—known collectively as rhinoceros beetles—relies on large, fallen trees for its development. Depending on the species, development can take anywhere from 9 months to 4 years! These species cannot survive, then, in forests that are selectively logged or subject to clearing of fallen dead wood. Young, secondary forests, with less variety in tree species and size, do not support the diversity of insect life found in older forests. Rhinoceros beetles also suffer from overzealous specimen collectors.

### Headlight Beetle (*Pyrophorus* sp.)

1.6 in. (4 cm). Has a bright, luminescent spot on each side of the shoulder. A spine on the underside of the head can be snapped to produce a violent click that thrusts the beetle into the air, useful both in evading predators and to right itself if it is turned over. Occurs on Caribbean and Pacific slopes at low and middle elevations. Feeds on rotting fruit and other plant matter. Active in the evening, just before total darkness; they are attracted to lights. Headlight Beetles are easy to see as they fly quickly through the forest: their paired spots produce a constant green or orange shine. Larvae are also bioluminescent. *Illustration not to scale.*

### Harlequin Longhorn Beetle (*Acrocinus longimanus*)

3 in (7.5 cm). Large. Distinctively patterned. Has extremely long front legs and antennae. Occurs throughout the country in low-elevation forests (ranges from Mexico to South America). Mainly diurnal but sometimes attracted to lights at night. Uses a variety of trees as hosts, usually selecting one with bracket fungi on the trunk. Female makes an incision in the bark and over 2 to 3 days lays about 20 eggs. The larvae bore into the wood, feeding for 7 months before pupating. Pupation lasts 4 months. Heavily infested trees usually die, but healthy trees are seldom attacked and so, unlike some other longhorn beetles, this species is not a serious pest.

## Order Megaloptera (Dobsonflies)

### Dobsonfly (*Corydalus* sp.)

Length (including wings): 5 in (12.7 cm). Large. Shiny, transparent wings held closed over back when resting. Male uses two long pincers for duelling with other males and to grasp female. The 11 species of dobsonfly in Costa Rica occur on both Caribbean and Pacific slopes, to 5,000 ft (1,525 m). Found along streams and waterways. Nocturnal, attracted to lights. Adults live for about 1 week; they do not usually feed, but may take sweet fluids if encountered. Females lay coin-sized chalky masses of 100 to 3,000 eggs on rocks near streams. Larvae drop into water on hatching and take at least a year to mature. Larvae can survive brief periods out of water using spiracles to breathe; on reaching maturity, they climb onto a rock or log to pupate, hatching about 10 days later. Larvae can deliver a painful bite.

## Order Hymenoptera (Wasps, Bees, and Ants)

### Paper Wasp (*Mischocyttarus* sp.)

1 in (2.5 cm). A medium-sized yellow-and-black wasp. Makes a small, open-celled nest located on branches in dense vegetation, under overhanging rocks, or on buildings. Female feeds on nectar and caterpillars, chewing the latter before serving them to her larvae. She also gathers wood fiber that is made into a paste for use in cell construction. One dominant female initiates nest construction and lays most of the eggs, placing a single egg in each open hexagonal cell. Worker females tend the eggs and larvae. Prior to pupation, the larva closes its cell with a silken cocoon, emerging after about a month. The cell may then be reused.

### Tarantula Hawk (*Pepsis* sp.)

2 in (5 cm). A very large wasp. Blue-black with contrasting rust-orange wings. Female sometimes seen frenetically searching the forest floor for tarantulas and other spiders; she hunts both male tarantulas that are searching for females and the females themselves. She first stings and paralyzes the tarantula, drags it to her burrow or to a premade nest, then lays a single egg on the body of the still-living tarantula. When the egg hatches, the larva feeds on the body fluids of the spider; as the larva develops, it tunnels into the host and feeds voraciously, avoiding the vital organs to keep the host alive as long as possible. Adult Tarantula Hawks, in less exotic fashion, feed on fruit and nectar. Male sits on vegetation, scanning for females. Female has a long stinger and can inflict one of the most painful stings known, but is not particularly aggressive toward humans.

**Orchid Bee** (*Euglossa cyanura*)

0.5 in (1.3 cm). Bright metallic green; length of tongue equals length of body. This and other species of orchid bee are common throughout CR at low and middle elevations. Found in rainforests. Male, an often seen visitor (and pollinator) of orchids, does not obtain nectar but instead harvests chemicals from the flower tissue that he stores in grooves in his hind legs. As he continues to visit orchids and collect chemicals, he becomes "drunk" and may be unable to fly for a period of time. Male later releases these volatile compounds during displays thought to attract females and increase mating success. Lone females are often seen gathering mud from trails that they use to construct cell nests in rodent burrows or on logs. Nests, built by a small group of females, contain 5 to 30 clay cells. The female supplies the cell with a pollen-and-nectar paste before laying a single large egg and sealing the cell with a clay lid. Female delivers a painful sting that is used in self-defense.

**Stingless Bee** (*Trigona fulviventris*)

0.2 in (0.6 cm). Small, black, with a slender orange abdomen. Occurs throughout the country at low and middle elevations. Found in both wet and dry forests. Built near the base of a large tree, the funnel-shaped entrance to the bee's underground nest is a sure indicator of its presence. At dawn, enormous numbers of bees pour out of the entrance, and a steady traffic of exiting and returning bees continues throughout the day. The underground nest houses up to 10,000 individuals. Worker bees forage for both nest-building materials (mud, fungi, feces, and plant resins) and food (pollen and nectar). On finding a rich food source, a bee returns to the nest, gathers up 1 to 10 recruits, and returns to the food. At sufficiently abundant food sources, the recruiting process continues until several hundred bees learn the location of the food.

## Acacia Ant (*Pseudomyrmex* sp.)

0.2 in (0.6 cm). Small, reddish ants (4 species of *Pseudomyrmex* occur in CR). Occurs on the Pacific slope at low elevations and also in the Central Valley. These ants live in the hollow thorns of the Bull-horn Acacia tree. A large colony may consist of 20,000 adults and 50,000 larvae. When a tree is disturbed, individuals release a pheromone that signals to large numbers of ants to rush out and sting cows, leaf-cutter ants, and other herbivores. The ants provide another service to the tree by clearing away seedlings and other plants from around the base of the trunk. In return, the Bull-horn Acacia supplies the ant with two types of food, Beltian bodies (protein-lipid nodules located at leaflet tips) and sweet nectar. Kiskadees, wrens, and orioles often nest in Bull-horn Acacia. The ants initially attack their nests but cease their attack—and leave the birds alone—once they become accustomed to the presence of the nests.

orange nodules are Beltian bodies

## Bullet Ant (*Paraponera clavata*)

Worker 0.8 in (2 cm). This is the largest ant in Central America. Dark maroon; large head; powerful mandibles. Occurs on Caribbean slope to 1,640 ft (500 m). Found in wet forests. Though sometimes seen during the day, the Bullet Ant is more active at night, when it forages on the ground and up into the canopy. Worker forages alone for insects, nectar, and other plant fluids, and returns to the nest with its booty. Nest, located underground at the base of a large tree, contains around 1,000 workers and about half as many immature forms. If the nest is attacked, workers deep within the nest move the brood away from the source of the attack, while workers near the surface rush out, releasing a musky odor and squeaking loudly. If the disturbance persists, they sting the attacker, imparting the most painful sting known according to the Schmidt Sting Pain Index—and the author's personal experience. Waves of intense, burning, throbbing pain can last 24 hours.

## Army Ant (*Eciton* sp.)

Soldier 0.6 in (1.4 cm). Six species of army ant occur in Costa Rica. Most species are reddish brown or black and have long legs. Individuals range in size from large, big-headed soldiers with hook-shaped mandibles to very small workers without enlarged heads or mandibles. Occur with great abundance in low-elevation rainforests, but also found in deciduous forests and cloud forests, to about 6,560 ft (2,000 m). Army ants are carnivores that feed on other ants, social wasps, and a wide range of other arthropods. Colony size ranges from 30,000 to over 1,000,000. At night they form bivouacs (clusters of hanging ants), often inside hollow trees or logs. At dawn the ants pour out in columns, forming a fan about 50 ft (15 m) in width. Prey caught early in the day are carried back to the bivouac; as the party advances, a new bivouac site is selected. Antbirds accompany army-ant swarms, eating insects that are stirred up by the ants. *Illustration not to scale.*

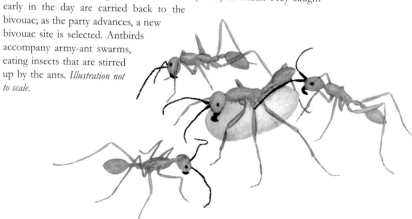

## Leaf-cutter Ant (*Atta cephalotes*)

Size varies greatly according to caste. Medium-sized worker ant (0.8 in/1 cm) is the most often seen; filing one after the other, worker ants carry pieces of leaves or flowers along well-groomed trails through the forest. Occurs throughout Costa Rica to about 6,560 ft (2,000 m). Underground nests are evidenced by extensive bare patches dotted with large entrance holes. Once the workers carry the leaves underground, smaller workers clean and scrape the leaves, then chew them into small pieces, adding saliva and fecal matter. This sticky mass is added to an existing fungus garden, which provides food for the colony. In a mature colony, hundreds of fungal gardens are connected by a labyrinth of tunnels. Without the addition of ant feces, the fungus does not produce spores and cannot grow. When a queen founds a new colony, she takes a piece of fungus with her, "fertilizing" it with her own eggs and feces before starting to raise new workers. *Illustration not to scale.*

**Order Araneae (Spiders)**

### Golden Orb-weaver (*Nephila clavipes*)

Female body length: 1 in (2.5 cm); male about a quarter the size of female. Brightly colored female is yellow and black and has yellow speckles. Male is a dull reddish brown. Usually found in clearings and second growth; hundreds of individuals may aggregate on telephone or electrical wires or under bridges. Female makes a web with a diameter of about 24 in (60 cm). Web, with characteristic sticky spiral near the center, is made from extremely strong silk that has been used to make gun sights and fishing lures. Upside down, the female waits for prey to become entangled. She bites the prey, injects venom, and, if the victim is large, envelops it with silk before consuming it. Several males may occupy a female's web. After mating, the female lays hundreds of eggs in a sac that she spins onto a tree.

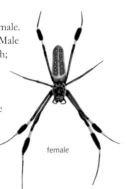

female

### Costa Rican Red Tarantula (*Brachypelma angustum*)

3.9 in (10 cm); leg span to 7 in (18 cm). Very large; blackish brown; long hairs on legs and abdomen. (Several species in the same genus occur throughout low- and middle-elevation forests of Costa Rica.) Constructs conspicuous burrows on the forest floor. Remains in burrow during the day, leaving at night to forage or search for mates. Feeds on insects and small vertebrates such as lizards, frogs, and mice. Very docile and seldom bites if handled; when defending itself, however, it kicks legs and the stinging hairs on these can cause a rash. Females lay a clutch of 100 to 600 eggs. Tarantulas are very slow-growing, reaching maturity only after 5 years; they live for up to 30 years.

## Bark Scorpion (*Centruroides limbatus*)

4.3 in (11 cm). Color varies from entirely blackish brown to yellow with black extremities. Long pincers are used to capture and hold food; tail is equipped with a stinger that causes a painful but not lethal sting. Occurs on Caribbean and Pacific slopes to 4,920 ft (1,500 m). Prefers forest edges of humid forests; also found near houses and other buildings. Nocturnal. At night, a UV light reveals the presence of scorpions; they contain a fluorescent compound and glow a bluish color. Eats spiders, insects, centipedes, and other scorpions. After a complex courtship, during which the male sometimes subdues the female by injecting her with a small amount of venom, the pair mate. The mother gives birth to 10 to 30 live young that crawl onto her back and remain with her for 2 weeks or more.

## Tailless Whipscorpion (*Paraphrynus laevifrons*)

0.8 in (2 cm). This bizarre creature resembles a cross between a hard-bodied spider and a tailless scorpion. The two front legs, very long and delicate, are used as sensory organs for locating prey. The clawlike mouthparts allow it to capture large invertebrates such as cockroaches. This species is mainly known from the Osa Peninsula, but it also occurs on the Caribbean slope to at least 3,280 ft (1000 m). During the day, rests in wells, burrows, or dugout pits; at night, often seen on the trunks of large trees. Female carries her egg sacs until they hatch, at which time the young climb onto her back. If any young fall off, the mother eats them.

## Large Forest-floor Millipede (*Nyssodesmus python*)

3.9 in (10 cm). The body of this large, conspicuous millipede is composed of flattish segments, with pale-cream keels attached to the side. Body segments vary in color. On some individuals, they are a uniform dark brown, on others a combination of brown and cream. Occurs on the Caribbean slope at low elevations. Very common at La Selva Biological Station. Feeds on rotting wood. When disturbed, it rolls up into a spiral and sprays from its hindgut a toxic liquid containing cyanide and benzaldehyde. This species mates venter to venter. After mating, the male "rides" the female for 5 days or more. If a second male attempts to mate with the female, the current "rider" flexes his body in order to force the female's head and anterior segments over her genital openings, thus preventing mating.

### Order Euonychophora (Velvet Worms)
This order belongs to a separate phylum (Onychophora). Members of this order are thought to be related to arthropods and tardigrades (water bears). Velvet worms, though not arthropods, are included here because of their resemblance to millipedes.

## Velvet Worm (*Epiperipatus isthmicola*)

3.1 in (8 cm). Wormlike. Stubby feet have claws. Occurs at low to middle elevations, in humid forests, second growth, and plantations, usually along streams. Uncommon and rarely seen; lives in leaf litter and remains out of sight even at night, when it is active. Slightly more abundant in areas with numerous ant nests, and may seek shelter in the nests. Predatory, taking insects, spiders, snails, and worms, including prey much larger than itself. It first squirts a sticky slime—to a distance of 12 in (30 cm)—to immobilize prey, then injects a toxic saliva that kills and predigests the victim.

**Order Decapoda (Crabs)**

### Halloween Crab (*Gecarcinus quadratus*)

Body width 3 in (7.6 cm). This colorful orange-and-black crab has purple claws. Common; occurs on Pacific slope, in mangroves and coastal rainforests. Very common on the Osa Peninsula, where it ventures inland to 1,970 ft (600 m). Feeds at night on leaf litter, seedlings, and, occasionally, animal food; drags food back to its burrow before consuming it. Sometimes climbs trees in search of food. It digs its own burrow, which can reach a depth of 5 ft (1.5 m). Adult lives in the forest but must return to the ocean to breed.

### Ghost Crab (*Ocypode* sp.)

Body width 2 in (5.1 cm). Adult are pale gray to reddish orange above; dull yellow or orange below; distinct purple to black eyes on long stalks. Young of most species are mottled gray-brown with mottled-gray eyestalks, blending perfectly with sand. The long eyestalks allow 360° vision. Occurs along the Caribbean coast (ranges from the United States to Brazil); occupies sandy beaches just above the tideline. This crab makes deep burrows in the dunes, and derives its name from its ability to disappear from view as it speedily retreats into a burrow. Large individuals sometimes steal the burrow of a smaller crab, and the displaced crab then takes over the burrow of an even smaller crab. Mainly active at night but also seen during the day. At dusk it scuttles to the ocean to obtain oxygen from the water, which it stores in sacs near the gills. Omnivorous.

### Hermit Crab (*Coenobita compressus*)

Shell 1.5 in (3.8 cm). This crab conceals itself—and lives—within an empty gastropod shell, and its size and appearance thus vary with the type of shell encountered. When they outgrow their shell, they search for a larger one. Occurs on Pacific coast beaches, from California to Chile. Mainly nocturnal but also active early mornings; during the day, shelters under logs and other hiding places. Varied diet includes detritus, vegetation, and feces of cows and horses; groups as large as 400 cluster around good food sources. Hermit crabs can be found far from the beach, but they require access to standing water, both for drinking and to replenish a supply that they carry in their shell. Female lays eggs in ocean, where her planktonic larvae develop.

# Soil Engineer

Halloween Crab (*Gecarcinus quadratus*).

Other common names for this crab, all befittingly colorful, are Moon Crab, Mouthless Crab, and the Harlequin Land Crab. Visitors to the Osa Peninsula and other rainforests on the Pacific coast of Costa Rica are guaranteed to see the Halloween Crab, particularly during the rainy season, when it is most active. In areas of high population density, this crab's burrows honeycomb the forest floor, and it becomes almost impossible to walk without stepping on one. The novelty—and delight—occasioned by the sight of large numbers of these crabs at a good distance from the ocean is a common experience for visitors to the country. Aside from the aesthetic pleasure they afford, however, Halloween Crabs play a critical ecological role. By carrying leaves into their burrows, they contribute to nutrient cycling; through selective eating of seeds they shape plant populations; and their burrows provide shelter for other arthropods.

# Glossary

**alkaloid** *n.* A type of active chemical compound found in many animal toxins.

**arboreal** *adj.* Living and feeding in trees.

**axillary** *n.* An area of feathers at the base of a bird's underwing that corresponds to the armpit of a mammal.

**carnivore** *n.* 1. A member of the order Carnivora. 2. Any meat-eating animal.

**cere** *n.* A fleshy covering over the nostrils of some birds.

**crepuscular** *adj.* Active at dawn or dusk.

**diastema** *n.* A space between teeth.

**disturbed area** *n.* Any habitat that—whether through human agency or natural forces—has been altered from its original state.

**dorsal** *adj.* Pertaining to the back of an animal.

**dorsum** *n.* The back of an animal's body.

**endemic species** *n.* Any species with a restricted geographic range. Note that the term is often used loosely, as in the phrase *endemic to Central America*, in which case the range of the animal may be relatively extensive.

**eyeshine.** *n.* Reflection from an animal's eyes when illuminated by flashlight at night. Usually best seen when the flashlight is held close to the eyes of the observer.

**folivore** *n.* An animal that specializes on eating leaves.

**forearm.** *n.* Arm from elbow to wrist; important in bat measurements.

**fossorial** *adj.* Having burrowing, secretive habits.

**frontal shield** *n.* An extension of the bill onto the forehead of some bird species.

**gallery forest** *n.* Evergreen forest along rivers.

**grizzled.** *adj.* Describes fur with a mix of dark and pale hairs or bands of color, from base to tip, on individual hairs.

**guard hair** *n.* One of the generally long hairs that extends beyond the undercoat of a mammal.

**harem.** *n.* A group of females defended by one male.

**keratin** *n.* The fibrous substance that forms horny tissues such as nails and hair.

**larva** *n.* The immature stage of an animal prior to metamorphosis. Frog and toad larvae are commonly referred to as tadpoles.

**lek** *n.* An assemblage of males in which the males vie for the attention of females through competitive vocal displays and choreographed dance performances.

**nape** *n.* The back of the neck.

**noseleaf** *n.* A leaf-shaped flap of skin above the nostril of some bat species.

**ovipositor** *n.* A female organ at the end of the abdomen that deposits eggs.

**oxidation pond** *n.* Open-air lagoon used to treat sewage.

**páramo** *n.* A cold, open, high-elevation habitat characterized by shrubs, grasses, and cushion plants.

**parotoid gland** *n.* A toxin-secreting gland near the ear in some toad species.

**parthenogenesis** *n.* The capacity of females of some species to produce offspring from an unfertilized egg. All offspring produced in this way are females.

**pelagic** *adj.* Living in the open sea.

**prehensile tail**. *n.* Tail that can be curled to grip a branch and that can support part or all of the animal's weight.

**primaries** *n.* The largest feathers on the edge of a bird's wing.

**sexual dimorphism** *n.* The systematic difference (in color, size, etc.) between males and females that occurs in some species.

**spiracle** *n.* A breathing vent found on some arthropods.

**syrinx** *n.* The vocal box of birds.

**tail covert** *n.* One of the downy feathers that cover the base of large tail feathers of a bird.

**venter** *n.* The abdomen or belly.

**ventral** *adj.* Pertaining to the underside of an animal; the belly.

# Further Reading

Forsyth, A., and K. Miyata. 1987. Tropical Nature: Life and Death in the Rain Forests of Central and South America. Touchstone, New York.

Gardner, A. L. (ed.). 2007. Mammals of South America. University of Chicago Press, Chicago.

Garrigues, R., and R. Dean. 2007. The Birds of Costa Rica: A Field Guide. Cornell University Press, Ithaca.

Hilty, S. 1994. Birds of Tropical America. Chapters Publishing, Shelburne.

Janzen, D. H. (ed.). 1983. Costa Rican Natural History. University of Chicago Press, Chicago.

Kricher, J. 1999. A Neotropical Companion. Princeton University Press, Princeton.

Kubicki, B. 2007. *Ranas de vidrio de Costa Rica* (Glass Frogs of Costa Rica). INBio, Costa Rica.

Leenders, T. 2001. A Guide to Amphibians and Reptiles of Costa Rica. Zona Tropical Publications, Costa Rica.

Reid, F. A. 2009. A Field Guide to the Mammals of Central America and Southeast Mexico. Second edition. Oxford University Press, New York.

Savage, J.M. 2002. The Amphibians and Reptiles of Costa Rica. University of Chicago Press, Chicago

Skutch, A. 1980. A Naturalist on a Tropical Farm. University of California Press, Berkeley.

Solórzano, A. 2004. *Serpientes de Costa Rica* (Snakes of Costa Rica). INBio, Costa Rica.

Stiles, F.G., and A. Skutch. A Guide to the Birds of Costa Rica. Cornell University Press, Ithaca.

Wainwright, M. 2007. The Mammals of Costa Rica. Cornell University Press, Ithaca.

Wilson, D. E., and D. M. Reeder (eds.). 2005. Mammal Species of the World: A Taxonomic and Geographic Reference. Third edition. Johns Hopkins Press, Baltimore.

# A Note on Amphibian Taxonomy

The species is the fundamental unit of biology, and each species of plant, animal, or microorganism possesses a unique and identifiable combination of morphological, genetic, and other biological traits. Closely related species are grouped together in the same genus, and related genera, in turn, are grouped within the same family. Biologists attempt to place every known living organism on the phylogenetic tree, which is commonly referred to as the *tree of life*, and the position of each species on this tree reflects its presumed relationship with other species.

As additional species are discovered and more sophisticated research tools become available, our insight into the relationships between species sometimes changes and the phylogenetic position of a species, genus, or family may require revision. In recent years, researchers have turned their attention toward several large families of amphibians, most notably Hylidae (Tree Frogs), Leptodactylidae (Rain Frogs and allies), Dendrobatidae (Poison-dart Frogs), Bufonidae (Toads), and Ranidae (True Frogs). In these studies, biologists have often for the first time analyzed a large number of individuals—as opposed to just a few specimens—and considered both morphological traits and molecular data as they grapple with questions of classification. The results have shaken the taxonomic foundations, and many large, unwieldy, and occasionally artificial, groupings of species have been broken up into smaller, more evolutionarily meaningful clusters. Since all of these newly formed groups need scientific names, many new genus and family names have emerged to describe Costa Rican amphibians.

An inherent problem is that a major taxonomic overhaul takes a while to completely catch on, and not all researchers agree with the findings of new studies, understandably enough, as there is always some degree of interpretation involved. Often, old and new names for a species exist side by side until one or the other is accepted by the scientific community, and even then some researchers may disagree and disregard proposed changes.

The phylogeny of Costa Rican amphibians is under intensive scrutiny at the moment, and it will be some time before the dust settles. Therefore, we chose not to include the most recently proposed taxonomic revisions, but use instead generally accepted, current names, many of which are themselves of fairly recent vintage. Since many of these new scientific names will be unfamiliar to the casual reader, and do not appear in older books, scientific names used prior to recent taxonomic changes are included in parentheses at the end of relevant species accounts, and, to avoid further confusion, newly *proposed* taxonomic changes are left unmentioned.

The table on the opposite page shows current taxonomy, in which names in blue indicate a relatively recent name change, and names in tan indicate proposed taxonomic changes.

| newest accepted changes | | | proposed changes | | |
|---|---|---|---|---|---|
| genus | species | family | genus | species | family |
| *Bufo* | *haematiticus* | haematiticus | *Rhaebo* | *haematiticus* | Bufonidae |
| *Bufo* | *marinus* | Bufonidae | *Chaunus* | *marinus* | Bufonidae |
| *Centrolene* | *prosoblepon* | Centrolenidae | *Centrolene* | *prosoblepon* | Centrolenidae |
| *Silverstoneia* | *flotator* | Dendrobatidae | *Silverstoneia* | *flotator* | Dendrobatidae |
| *Oophaga* | *pumilio* | Dendrobatidae | *Oophaga* | *pumilio* | Dendrobatidae |
| *Hylomantis* | *lemur* | Hylidae | *Hylomantis* | *lemur* | Hylidae |
| *Hypsiboas* | *rosenbergi* | Hylidae | *Hypsiboas* | *rosenbergi* | Hylidae |
| *Hypsiboas* | *rufitelus* | Hylidae | *Hypsiboas* | *rufitelus* | Hylidae |
| *Isthmohyla* | *pseudopuma* | Hylidae | *Isthmohyla* | *pseudopuma* | Hylidae |
| *Dendropsophus* | *ebraccatus* | Hylidae | *Dendropsophus* | *ebraccatus* | Hylidae |
| *Dendropsophus* | *microcephalus* | Hylidae | *Dendropsophus* | *microcephalus* | Hylidae |
| *Scinax* | *elaeochrous* | Hylidae | *Scinax* | *elaeochrous* | Hylidae |
| *Trachycephalus* | *venulosus* | Hylidae | *Trachycephalus* | *venulosus* | Hylidae |
| *Gastrotheca* | *cornuta* | Leptodactylidae | *Gastrotheca* | *cornuta* | Hemiphractidae |
| *Craugastor* | *bransfordii* | Leptodactylidae | *Craugastor* | *bransfordii* | Craugastoridae |
| *Craugastor* | *fitzingerii* | Leptodactylidae | *Craugastor* | *fitzingerii* | Craugastoridae |
| *Craugastor* | *megacephalus* | Leptodactylidae | *Craugastor* | *megacephalus* | Craugastoridae |
| *Eleutherodactylus* | *diastema* | Leptodactylidae | *Diasporus* | *diastema* | Eleutherodactylidae |
| *Eleutherodactylus* | *ridens* | Leptodactylidae | *Pristimantis* | *ridens* | Strabomantidae |
| *Leptodactylus* | *savagei* | Leptodactylidae | *Leptodactylus* | *savagei* | Leptodactylidae |
| *Engystomops* | *pustulosus* | Leptodactylidae | *Engystomops* | *pustulosus* | Leiuperidae |
| *Rana* | *vaillanti* | Ranidae | *Lithobates* | *vaillanti* | Ranidae |
| *Rana* | *vibicaria* | Ranidae | *Lithobates* | *vibicarius* | Ranidae |
| *Rana* | *warszewitschii* | Ranidae | *Lithobates* | *warszewitschii* | Ranidae |

# Index of Scientific and Common Names

## Mammals

## Reptiles

## Amphibians

# Arthropods

# About the Authors

**Fiona A. Reid** has a BA degree in Biology from Cambridge University and an MSc in Animal Behavior from SUNY at Stony Brook, New York. She has written and/or illustrated more than a dozen books on mammals, including *A Field Guide to the Mammals of Central America and Southeast Mexico* (Oxford University Press, second edition, 2009). While researching that book she lived in Central America for two years, capturing small mammals and drawing them from life. Fiona also wrote and illustrated a *Peterson Field Guide to Mammals of North America,* published in 2006 by Houghton-Mifflin. Fiona is a Departmental Associate of the Centre for Biodiversity and Conservation Biology at the Royal Ontario Museum in Toronto.

**Twan Leenders** was born and raised in The Netherlands, where he completed his doctoral exam in Biology, specializing in Animal Ecology. Since the early 1990s, he has worked (and lived) in various countries in Central America and tropical Africa to study amphibians and reptiles. In recent years, his research has focused on the precipitous decline in amphibian species and the dynamics of surviving populations. After moving to the United States in 2001, he worked in the Department of Vertebrate Zoology of Yale University's Peabody Museum of Natural History before joining The Connecticut Audubon Society in 2008 as the organization's conservation biologist.

**Jim Zook** is an ornithologist who has lived and worked in Costa Rica since 1988. He first came to the country as a Peace Corps volunteer to teach environmental education. A native of Ohio, he holds a BA degree in Zoology from Colorado State University. He lives with his wife and son in Naranjo de Alajuela.

**Robert Dean** has been studying and painting neotropical birds for twelve years, during which time he has been on birding trips throughout the Americas, both as guide and as tour participant. Born and raised in London, England, he was a professional musician for eighteen years before moving to Costa Rica, where he revitalized his childhood passion for wildlife and art. Previous accomplishments include his artwork for *The Birds of Costa Rica: A Field Guide* (Cornell University Press 2007); he is currently painting illustrations for an upcoming field guide to the birds of Panama.